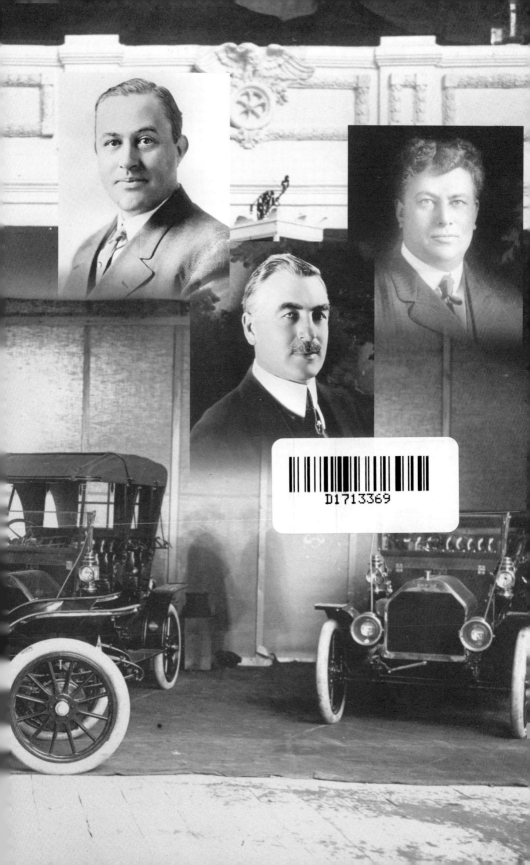

The E-M-F Company

The Story of Automotive Pioneers Barney Everitt, William Metzger, and Walter Flanders

WITHDRAWN

Prepared under the auspices of the
SAE Historical Committee

Other books in the SAE Historic Motor Car Company Series:

**The Franklin Automobile Company:
The History of the Innovative Firm, Its Founders,
The Vehicles It Produced (1902-1934),
and The People Who Built Them**
by Sinclair Powell
(Order No. R-208)

For more information or to order this book, contact SAE at 400 Commonwealth Drive, Warrendale, PA 15096-0001; phone (724) 776-4970; fax (724) 776-0790; e-mail: publications@sae.org.

The E-M-F Company

The Story of Automotive Pioneers Barney Everitt, William Metzger, and Walter Flanders

Anthony J. Yanik

Society of Automotive Engineers, Inc.
Warrendale, Pa.

Library of Congress Cataloging-in-Publication Data

Yanik, Anthony J.
 The E-M-F Company : the story of automotive pioneers Barney Everitt, William Metzger, and Walter Flanders / Anthony J. Yanik.
    p. cm. — (Historic motor car company series)
 Includes bibliographical references and index.
 ISBN 0-7680-0716-X
 1. Everitt, Barney. 2. Metzger, William. 3. Flanders, Walter E. (Walter Emmett), 1871–1923. 4. E-M-F Company—History. 5. Automobile engineers—United States—Biography. I. Title. II. Series.

TL140.E94 Y35 2001
629.2'092'273—dc21
[B]   2001020707

Copyright © 2001 Society of Automotive Engineers, Inc.
            400 Commonwealth Drive
            Warrendale, PA 15096-0001 U.S.A.
            Phone: (724) 776-4841
            Fax: (724) 776-5760
            E-mail: publications@sae.org
            http://www.sae.org

ISBN 0-7680-0716-X

All rights reserved. Printed in the United States of America.

Permission to photocopy for internal or personal use, or the internal or personal use of specific clients, is granted by SAE for libraries and other users registered with the Copyright Clearance Center (CCC), provided that the base fee of $.50 per page is paid directly to CCC, 222 Rosewood Dr., Danvers, MA 01923. Special requests should be addressed to the SAE Publications Group. 0-7680-0716-X/01-$.50.

SAE Order No. R-286

Contents

Preface		vii
Introduction	Setting the Scene	ix
Timeline		xi
Chapter One	From Carriages to Car Bodies— Byron F. "Barney" Everitt: His Early Years	1
Chapter Two	The Making of Cadillac and Other Daring Deals— William E. "Bill" Metzger: His Early Years	11
Chapter Three	The Merry Master of Mass Production— Walter E. Flanders: His Early Years	35
Chapter Four	EMF Bursts onto the Automotive Scene	45
Chapter Five	EMF Loses Its "E" And "M"	75
Chapter Six	Flanders Expands EMF, Declares War on Studebaker	89
Chapter Seven	Crisis or Comedy? Studebaker Sues EMF	107
Chapter Eight	The Rebirth of Everitt and Metzger— Flanders, Where Are You?	143
Chapter Nine	Flanders Reunites with Everitt and Metzger	155

Chapter Ten   "E" and "M" and "F" After 1913—
               The Dance Continues 173

Epilogue 205

Endnotes 207

Index 225

About the Author 243

Preface

I became fascinated with the accomplishments of Messrs. Everitt, Metzger, and Flanders several years ago while researching the early days of the auto industry in Detroit. What intrigued me was the frequency with which their names kept recurring in the automotive press between 1900 and World War I, if not beyond. It soon became obvious that Barney Everitt, Bill Metzger, and Walter Flanders were as highly regarded by their peers as Henry Ford, Will Durant, Henry Leland, or R.E. Olds. However, today Everitt, Metzger, and Flanders are virtually unknown. For reasons that escape me, contemporary historians more often than not simply ignore them.

However, in 1908, when they founded the E-M-F Company (EMF), Everitt, Metzger, and Flanders were known on the streets of Detroit as the "Big Three" of the auto industry. Within three years, EMF was the largest employer in Detroit and was producing more cars than any other company in the United States other than Ford. Were it not for the fact that the Studebaker Brothers Manufacturing Company purchased the entire E-M-F Company for an outrageous price in 1910, EMF might still be an auto giant today. On the shoulders of EMF, Studebaker (until then, the largest wagon builder in the country) built its strong automotive base. Before the purchase of EMF, Studebaker had been only a bit player within that industry.

Barney Everitt, Bill Metzger, and Walter Flanders were automotive pioneers in many other ways outside of EMF. Everitt was instrumental in forming the extensive body building industry that characterized Detroit prior to World War II. Metzger in turn established the first automobile dealership in Detroit, if not the country. As head of sales for Cadillac, he virtually guaranteed the success of that company in its formative years. Flanders, a genius with machines, masterminded the tools of production for the first Model T.

By documenting the careers of Everitt, Metzger, and Flanders, I have provided some insight into the typical wheeling and dealing that went into the formation of automobile companies in Detroit prior to World War I. It truly was a freewheeling era during which men of great skill, courage, drive, and fortitude overcame a lack of formal education to create wealth for themselves and for those willing to back their efforts.

For easier reading, I have taken the liberty of substituting the terms EMF or the EMF Company for the more correct E-M-F Company or the Everitt-Metzger-Flanders Company. I also generally refer to their automobiles, advertised as the E.M.F. "30" and the Flanders "20," as the EMF 30 and the Flanders 20. The fact that both the company and the car used E-M-F in their logo lent itself to much friendly satire. One of the most popular of these was "Every Mechanical Fault."

For the information that went into the writing of this history, I owe a debt of gratitude to Mr. and Mrs. Robert Denham of the Studebaker National Museum, who allowed me to prowl through the EMF board minutes and provided me with copies of pages without which this account would have been impossible to write. My thanks also are extended to Mark Patrick, curator of the National Automotive History Collection (NAHC) of the Detroit Public Library, who provided the photographs that appear in this book and made available to me the NAHC collection of early auto periodicals, as did the late Richard Scharchburg and Bill Holloran of the Collection of Industrial History at Kettering University. In addition, I am grateful to Barbara Fronczak and the Chrysler Archives, where the board minutes of the Metzger, the Everitt, and the Flanders Motor Car Companies reside, and to the State Archives of Michigan, the source of annual reports of old auto companies.

Introduction
Setting the Scene

The year is 1908.

Before 12 months have passed, 1908 will have become a momentous year of change for the auto industry.

This change first is signaled in February when a Thomas Flyer wins the grueling New York to Paris race, thus putting to rest the vaunted superiority claimed by European automakers over U.S.-built cars.

A few weeks later, officials of the Royal Automobile Club disassemble three Cadillac cars, mix their parts together, and then rebuild the cars. The three vehicles are taken to a track and, to everyone's amazement, are driven for 500 miles without a hitch—proving that automobile manufacturing techniques within the United States have reached a stage of standardization unequalled by any other industrial nation.

In September, the automotive world is startled by the creation of the first successful conglomerate in the auto industry: General Motors Corporation, a product of the visionary outlook of Will Durant and Ben Briscoe.

At the beginning of the next month, October 1, a homely and unpretentious new Ford model emerges from the factory on the corner of Piquette Avenue and Beaubien Street in Detroit. It is called the Model T, the ninth in the series of cars built thus far by Ford. Demand for the Model T will become so great that production soon will cease for nine weeks to enable orders to be filled. More than 15 million Model T's will be built during the next 19 years.

Several months before the first Model T is produced, the factory manager responsible for setting up its initial production—a man of genius with machines—leaves Ford to join with two others to form their own company. His name is Walter Flanders.

Mark it well.

Joining Flanders are Byron F. "Barney" Everitt, pioneer body builder and trimmer, and William E. "Bill" Metzger, the first auto dealer in Detroit and perhaps the nation. Each of the three men is well known within the auto industry of this time—as well known as Henry Ford, Ransom Olds, Henry Leland, Will Durant, Ben Briscoe, and the Dodge brothers. They are movers and shakers. When they speak, the automotive world listens and gives them deference, so strong have their reputations become within the industry.

On June 2, 1908, the three men call a press conference in New York City to announce that they have formed the Everitt-Metzger-Flanders Company (EMF) and will begin production of a medium-priced car in Detroit in September. Oddly enough, the EMF factory will be located on the next block west of the Ford plant.

That three such powerful automotive names have come together to build a new automobile is sensational news. Proof of the magic of their names will come from sales figures as the months unfold. By the close of 1909, EMF will vault into the Number Two slot in automotive production—truly a remarkable achievement in such a short period of time. Part of the success of the new company will be due to its alliance with the Studebaker Brothers Manufacturing Company, then the largest wagon producer in the nation. The Studebaker sales outlets will be used to market EMF automobiles, but the association is acrimonious. Eventually, Studebaker will buy out EMF and use it as the foundation for a reorganized Studebaker Corporation that will become a leader among second-tier auto producers (behind Chrysler, General Motors, and Ford) for almost half a century.

This is the story of Barney Everitt, Bill Metzger, Walter Flanders, and the EMF Company, which augured so much success during the first decade of the twentieth century. Had the three men not sold out to Studebaker, it is conceivable that their names would be familiar to us as founders of a large automobile company that continues to exist to this day.

Timeline

1908

- **June 2** — Formation of Everitt-Metzger-Flanders (EMF) Company announced.
- **August 4** — EMF incorporated in Michigan. Everitt voted president, Metzger secretary, and Flanders general manager.
- **August 5** — EMF signs sales agreement with Studebaker Brothers Manufacturing Company.
- **September 15** — EMF purchases Wayne Automobile Company.
- **October 5** — EMF purchases Northern Motor Car Company.

1909

- **March 4** — Everitt's letter to EMF board condemns Studebaker sales agreement.
- **April 12** — Flanders' letter to EMF board cites need for Studebaker.
- **April 21** — EMF board proposes new sales agreement to Studebaker.
- **April 29** — Everitt and Metzger resign after Studebaker buys their EMF stock.
  William Kelly resigns.
  Flanders appointed president.
  EMF signs new sales contract with Studebaker per latter's terms.

1909

July 19	EMF buys De Luxe Motor Car Company.
July 27	EMF announces purchase of Western Malleable Steel & Forge Co. and Monroe Body Company.
December 9	Flanders' letter to Studebaker terminates EMF/Studebaker sales agreement.
December 11	EMF advertises for sales agents to replace Studebaker.
December 13	Studebaker files bill of complaint against EMF in Detroit (Judge Swan).
December 14	Detroit court (Judge Swan) refuses to grant injunction preventing EMF from selling its own cars.
December 16	Studebaker files second bill of complaint against EMF in Cincinnati (Judge Severens) and requests injunction to prevent EMF sales.
December 17	Cincinnati court (Judge Severens) informs EMF of injunction halting EMF car sales.
December 24	Cincinnati court (Judge Severens) lifts injunction against EMF and drops Studebaker suit.
December 27	Studebaker asks Detroit court (Judge Swan) to withdraw its original suit.
December 29	Detroit court (Judge Swan) agrees to drop Studebaker suit.
December 30	Studebaker files new bill of complaint in Kalamazoo (Judge Severens) asking for injunction to stop EMF car sales.
	Studebaker also files another suit in Cincinnati (Judge Warrington) asking for similar injunction.
December 31	At special EMF board meeting, majority of directors vote support of Flanders' letter rescinding Studebaker sales contract.
	Same day, EMF asks Wayne County court to remove Studebaker members from EMF board.

1910

- **January 3** — Studebaker files new suit (Judge Swan) against EMF in Detroit asking for damages and injunction to prevent further sales by EMF of its cars.
- **January 10** — Cincinnati court (Judge Warrington) refuses Studebaker complaint filed on December 31 and apologizes for wasting court's time.
- **February 10** — Detroit court (Judge Swan) dismisses Studebaker suit filed on January 3.
- **March 5** — Studebaker drops any suits still pending against EMF.
- **March 8** — J.P. Morgan & Company purchases EMF for Studebaker.

  New officers are elected, with Flanders appointed president and general manager.

1911

- **February 14** — Studebaker Brothers Manufacturing Company reorganizes as Studebaker Corporation as a result of Fish being forced to borrow new money to pay the notes due for purchasing EMF.

  EMF now becomes "auto division" of Studebaker Corporation.
- **April 24** — Flanders agrees to new contract as general manager of automotive operations, on condition that Studebaker purchases his stock in the Flanders Manufacturing Company.

1912

• August 7 — Flanders resigns from Studebaker and rejoins Everitt and Metzger as general manager of Everitt Motor Car Company (originally named Metzger Motor Car Company) which Everitt and Metzger had formed after leaving EMF. Subsequently, Everitt Motor Car Company is renamed Flanders Motor Car Company in August 1912 and eventually is purchased by Maxwell Motor Company in April 1913.

Chapter One

From Carriages to Car Bodies—
Byron F. "Barney" Everitt: His Early Years

It was the age of Horatio Alger, a mythical time when a person could rise from rags to riches by the sheer dint of hard work, and only in America. However, it was not always myth, as events so often proved during the early days of the automobile. For those who dared, the myth proved that sometimes it could become a reality.

One man who dared was Byron F. "Barney" Everitt.

Everitt at age 36 became head of the Wayne Automobile Company, the first building block that eventually would result in the creation of the EMF Company in 1908. In time, EMF would present a strong challenge to the rest of the auto industry for sales supremacy and lose its identity only after being purchased by the Studebaker Corporation.

Everitt, the Carriage Body Apprentice

Everitt was born in Ridgetown, Province of Ontario, Canada, on May 3, 1872. He decided at an early age that he was not meant to spend his young days behind a school desk. He went off to Chatham, Ontario, about 80 miles east of Detroit, where he entered the employ of William Gray & Son, one of the best-known Canadian carriage makers of that period. Everitt worked hard at

Byron F. "Barney" Everitt in his prime. (Courtesy of the Detroit Public Library, National Automotive History Collection)

mastering the trade. When he reached the lofty age of 19, he left Gray's employ and journeyed down the road and across the river to Detroit. No doubt with Gray's recommendation (Gray considered Everitt to be one of the best men who had ever worked for him), he had no difficulty finding employment with Hugh Johnson, another veteran carriage builder in the Detroit area.[1-1]

Again, Everitt quickly gained a reputation for mastering the skills of his trade. In addition, talk was circulating among those in the business about Everitt's growing aptitude for management. Soon the latter became more than merely talk. Within two years of joining the Hugh Johnson concern, Everitt in 1893 received a most attractive offer from the C.R. Wilson Carriage Company to run the trimming and finishing departments of that company.

This indeed was an offer not to be ignored. Eventually, C.R. Wilson would claim to be the largest carriage and buggy builder in the world.[1-2] This claim, of course, was open to criticism. Nevertheless, the offer to Everitt substantially upgraded his status within the carriage profession. It would have other ramifications as well because the C.R. Wilson Carriage Company adopted a favorable attitude toward the embryonic Detroit auto industry. It was an outlook no doubt passed on to young Barney Everitt—one that would reward him handsomely.

In the fall of 1899, three years after having married the attractive Donna Shinnick, Everitt made his leap toward independence by opening his own carriage trimming business on the corner of Brush and Woodbridge streets in downtown Detroit, only a few blocks south of William Metzger's automobile agency. Although carriage trimming initially was the name of his game, Everitt was thoroughly trained in the art of carriage building and had no problem expanding into the latter area when the occasion demanded.

That occasion made itself known rather quickly.

Mr. Olds, At Your Service

In the same year that Barney Everitt set up shop in downtown Detroit, incorporation papers were filed with the state of Michigan for the Olds Motor Vehicle Company. Chief financial backer for Olds was Samuel L. Smith from Detroit. Thus, it was no surprise when Olds announced that it had purchased five acres of land along the Detroit River near Belle Isle as the site for a new manufacturing facility. The buildings were completed and production was underway by the first of the year in 1900. Earlier, Ransom Olds had contracted with Everitt to build the bodies for all Olds models produced in Detroit.[1-3] This may not necessarily be true. Indications are that C.R. Wilson built bodies for Olds as well.[1-4] To further tangle matters, C.B. Glasscock, in his book, *The Gasoline Age,* writes that Everitt was the person who invited C.R. Wilson to join him in supplying Olds bodies because his plant was unable to keep up with all the requests coming from the Olds Motor Works. Because C.R. Wilson also had difficulty keeping up with the pace, Everitt journeyed to Norwalk, Ohio, to convince two young fellows whom he had previously met to come north and give them a hand. Their names were Fred and Charles Fisher, two of seven brothers working in their father's carriage and blacksmith shop.[1-5] The two brothers eventually parlayed their entry into the Detroit automobile industry into their own Fisher Body Company, probably the most well-known body company to develop in these early, halcyon days of automobile development.

On March 9, 1901, at 1:35 in the afternoon, a disastrous fire swept through the Olds complex. Fortunately, the work force had been sent home for the day. Only 24 men were on hand, and they escaped. The entire factory except the foundry was seriously damaged. However, all the drawings and plans of existing and future models were found intact within the vault in the main building. Ransom Olds immediately announced that the Olds Motor Works had on hand sufficient orders to produce vehicles for the remainder of the year and would set up temporary production facilities as quickly as possible.[1-6]

The fire often has given rise to the myth that one vehicle was saved—an experimental curved dash runabout that then became the prototype for future Olds production, the success of which subsequently made Ransom Olds famous.

Although the curved dash Olds was the only one of twenty vehicles to survive the fire, ads touting the runabout had been published a month before the fire. Because no other vehicle types were mentioned in that ad, it is conceivable that Olds already had decided to make the curved dash runabout the centerpiece of the Olds Motor Works. The fire only delayed production.[1-7]

More importantly, the fire gave Ransom Olds a reason to move production facilities back to his roots in Lansing. Although the Detroit plant eventually was rebuilt and shared curved dash production with Lansing, Olds remained in Lansing rather than divide his time between the two factories. This became a sore spot between him and the Detroit-based Smith, which ultimately led to Olds' departure from his own company.

Everitt and Wilson continued to supply bodies for the Detroit Olds plant. However, as the Lansing branch began to take precedence, Everitt ultimately was phased out of the supply chain, which was just as well.

Henry Ford Beckons

A new opportunity was presenting itself to Everitt in Detroit in the fall of 1902. A young man who was making quite a name for himself within Detroit auto racing circles, Henry Ford, had entered into a partnership with Alexander T. Malcolmson, a Detroit coal dealer, to produce a prototype of an automobile which they then could shop around for investment money. By November 1902, Ford could be found hammering out angle iron in his shop, preparatory to assembling a chassis. Once finished, he needed a body on which to mount it. Without hesitation, he approached Everitt to supply that same body. It seemed a logical choice because Everitt by then had become the largest supplier of body trim in the country, and his shop on Fort Street was one of the best equipped for such work in the land.[1-8]

At this point, Everitt's association with the Ford Motor Company becomes sketchy. One account claims he built bodies for the first 10,000 automobiles produced by Ford.[1-9] If this indeed was the case, Everitt was the sole supplier of Ford bodies between 1903 and about 1907 because that was the approximate total Ford production of complete vehicles over these four years.

By 1906, Everitt had moved into new quarters at Clay Avenue and the Grand Trunk Railroad.[1-10] Actually, the facility was located on the southwest corner of Clay and Dequindre avenues. The significance of the move was that it placed the Everitt works several miles north of downtown Detroit at a location not too distant from the doorstep of the Ford Motor Company on Piquette Avenue (both of which were near the site of the future General Motors Building). Ford had relocated to the Piquette Avenue address in 1905.

Walter O. Briggs Comes On Board

In 1904, Everitt made a significant addition to his company when he lured Walter O. Briggs away from the C.H. Little Company by offering to make Briggs the general manager of the Everitt works.

Briggs was another one of those self-made Horatio Alger-types we encounter so often in the early days of the auto industry. Born in Ypsilanti, Michigan in 1877, his father a locomotive engineer, Briggs went to work as a car checker or yardman with the Michigan Central Railroad when he was only 14 years old. He gradually made his way upward through the ranks to foreman, then left Michigan Central for Detroit and the C.H. Little Company in 1902, where he became foreman of the paint and trim shop.[1-11] Briggs must have made quite an impression on his peers for Everitt to select him to manage the Everitt enterprise in 1904. However, Everitt had good reason to groom someone to take his place in running the business, for it appears that he was about to take a new fork in the road of his business life.

The Wayne Adventure: On the Road to EMF

On November 10, 1904, a new automobile concern was incorporated in Detroit called the Wayne Automobile Company. The company had been formed about a year earlier and had quickly made a name for itself. As reported during the spring of 1904 in *The Detroit Journal*,[1-12]

> While young, the business is pushing ahead rapidly and bids fair to soon become one of Detroit's foremost automobile industries.

At what point Barney Everitt became a vital cog in the fortunes of the company is not clear, but Wayne was to have a powerful effect on Everitt's future within the automotive industry. One reference mentions Everitt as being general manager of the Wayne Automobile Company for a number of years.[1-13] Therefore, it is conceivable to assume that Everitt was offered this position when the company was formed, although his name is not mentioned prominently at that time.

In any event, if bets had been taken among Detroiters of that time in predicting which auto firms then operating in Detroit would be successful, Wayne would have been at the top of the list with the likes of Ford, Cadillac, and Packard. The Wayne roster included two of the most well-known names in Detroit's elite society: Charles Louis Palms, listed as both president and treasurer, and James Book, one of its five directors. Moreover, Palms and Book were amply endowed financially.

Palms and Book: Early Detroit Auto Magnates

At only 33 years of age, Charles Palms was a man of the arts but entirely at home when dealing with financial matters. His grandfather, Francis Palms, was considered to be the largest landowner in Michigan during his lifetime. He had vast holdings of timber and minerals in upper Michigan, Wisconsin, and Ontario, Canada, as well as real estate throughout the Detroit area. He also had invested heavily in the Michigan Stove Company, serving as its president from January 1873 until his death in 1886. Charles' father, Francis F. Palms, who had settled in New Orleans where Charles was born in 1871, subsequently moved back to Detroit after grandfather Palms died, to take over the management of the family estate.[1-14]

Meanwhile, Charles had graduated from Georgetown University with a Ph.D. at the age of 18, studied law at Harvard University, aspired to be a journalist, and had become proficient at playing the violoncello—all by the time he had reached the age of 21. Unfortunately, his health began to suffer from all these pursuits, and his father enticed him to come to Detroit to replace him in managing the Francis Palms estate. The Palms estate probably was worth approximately $21 million when Charles took over its management.[1-15] He soon whipped the estate into such good shape that he was free to turn to other business ventures. Subsequently, Charles was elected president of the Preston National Bank in 1901 and that same year became secretary and treasurer of *The Detroit Journal,* a leading newspaper of that era.[1-16] The Palms family also was represented on the boards of the two largest stove companies in Detroit—the Michigan Stove Company and the Detroit Stove Works.

James Book, Charles' uncle by marriage, had taken a different route to success. One of the leading citizens of Detroit, although a Canadian by birth, Book graduated from the Jefferson Medical School in Philadelphia when he was only 21 years old and soon migrated to the Detroit area. By 1872, still not 30 years old, Book had become chief surgeon of the Detroit, Lansing & Northern Railroad and medical director of the Imperial Life Insurance Company. Four years later, he was appointed head of the surgical staff of Harper Hospital in Detroit and had developed an enviable medical reputation for the many papers he had published on his original experiments. Independently wealthy, Book retired from the medical profession in 1895 to begin a new occupation—that of being a capitalist. By the time the Wayne Automobile Company opportunity had come his way, Book already was the director of two banks and had investments in both the Anderson Carriage Company and the Anderson Electric Car Company.[1-17]

We would expect that the Wayne Automobile Company, with such financial firepower behind it, would have little difficulty in seeing its way through the thickets of the auto industry. Its capitalization represented $300,000 in common stock, of which 265,000 shares were subscribed. Palms held 75 shares and was trustee for another 225. Edward A. Skae, a local coal dealer, owned 75 shares; Roger J. Sullivan held 1,133 shares; and William Kelly held 1,132 shares.[1-18] Wealthy James Book represented only 10 shares. Ironically, almost nine of ten shares were held by the two men directly involved in the production of the Wayne automobile—Sullivan and Kelly. With Palms being president, we can only assume that the Sullivan and Kelly shares were non-voting shares. That is, if the business failed, they were out the time and energy that they had put into the firm. If Wayne became a success, the rise in the value of their stock would be their reward.

The Wayne Car

Contemporaries suggested a bright future for the Wayne automobile. The initial offering of the company in 1904 was a two-cylinder, 16 horsepower touring car, the Model A. Designed by William Kelly, it was a sturdy (if not spectacular) automobile that sold at a midrange price of $1,200. For 1905, Kelly engineered a larger, four-cylinder, 24 horsepower Model C that listed for $2,000.

By 1906, the company was offering six different models: three at two cylinders, and three at four cylinders. The bottom of the line was the Model H, a two-cylinder, 14 horsepower runabout that sold for only $800. The top of the line was the Model F, a four-cylinder, 50 horsepower model stretched out on a 117-inch wheelbase and selling for $3,500.[1-19] The latter decidedly was a luxury car, no doubt one in which Charles Palms would prefer to be seen riding on his way to the office. It was a rather ambitious lineup of vehicles for a small concern, offering a greater variety than that of any of the leading auto manufacturers of that year, including Ford, Cadillac, and Buick. An interesting item in one of the Wayne advertisements claimed that "the entire engine can be taken apart or assembled in half an hour."

This 1906 Wayne Model H runabout, shown here in restored condition, originally sold for $800.

Evidence seemed to indicate that a new wave of excitement was building within the Wayne Automobile Company during 1906.[1-20] There certainly was evidence of several noteworthy changes that could be gleaned from the annual report of the company, dated December 31, 1906. First, the home address of its operations now was given as the corner of Brush Street and Piquette Avenue on the then northern edge of Detroit. There is nothing outwardly significant about a company moving into new quarters, but this move located the new Wayne plant one block away from the factory of Henry Ford. As will be seen later, this may have had significant bearing on the future of the Wayne enterprise.

Also, two new names were added to the roster of Wayne stockholders: George Wilson from Indianapolis, and, more importantly, Dona F. Everitt, wife of Barney Everitt.[1-21] Unfortunately, the annual report does not give any indication of the number of shares held by each stockholder, although the total number subscribed had been increased by seven. The roster of officers and directors continued unchanged from when the company was formed.

Wayne in Trouble: Everitt Takes Over

With debts totaling $2.7 million against assets of nearly $5.5 million, the Wayne Automobile Company seemed to be in good shape for the coming year of 1907. Apparently, this was not the case, despite the fact that Wayne would produce 1,000 units that year, which was not too bad for those times. Whether the national financial crisis of 1907 affected the company is not known, but Charles Palms had been sufficiently disturbed by poor earnings to remove himself from the presidency. Part of the problem appeared to be the company's approach to building a car, which was not much different from the already antiquated process of stock runners bringing materials to a spot on the floor where a single car was assembled under the direction of a master mechanic. It also could be that Palms brought the problem onto himself by shifting the production of the Wayne Automobile Company to that of only large high-priced luxury touring cars ranging in price from $2,500 to $3,500.[1-22]

Palms' choice as new president was none other than Barney Everitt, who took office during September 1907. Thus, the first leg was attached to the three-legged stool that in the summer of 1908 would become known as the EMF Company.

Chapter Two

The Making of Cadillac and Other Daring Deals—

William E. "Bill" Metzger: His Early Years

If ever a man earned the name of "Mr. Motorcar" during the first decade of the automobile industry in Detroit, that man would be William E. "Bill" Metzger. Metzger "invented" the auto dealership, gave birth to Detroit's first automobile show, and helped form five automobile companies, among his many other ventures. He was the foremost salesman and promoter of the industry during those halcyon days when Detroit was establishing itself as the center of the passenger car trade.

Metzger, The Two-Wheeler Dealer

Metzger was born in Peru, Illinois, on September 30, 1868. His father, Ernest, was a German immigrant who served in the Civil War; his mother, Maria, was a native of Ohio. At some point, the family must have moved to Detroit because Bill Metzger officially graduated from the Detroit public high school system in 1884 when he was 16 years old.[2-1] He immediately went to work for a furniture store titled Hudson & Symington where his father held an important position. The man who operated the store was J.L. Hudson, who later opened what eventually became one of the leading department stores in the United States, a popular fixture in downtown Detroit for generations.

The E-M-F Company

William Metzger, the nation's first auto dealer. (Courtesy of the Detroit Public Library, National Automotive History Collection)

As a young man, Metzger, similar to so many others of that era, was caught up in the rising tide of enthusiasm for the bicycle which then was sweeping across the country. He joined and was the first president of the Detroit Wheelman's Club, a branch of the League of American Wheelmen, and he earned a number of century badges which were awards given to bicycle riders who could complete 100 miles within a single day.[2-2] His early bicycle racing was done on the high-wheel bone-crusher type of bicycle typical of the time.

The year after Metzger graduated from high school, a man named James Starley introduced the modern multiple-geared, low-wheeled bicycle to the world. This and the invention of the pneumatic tire a few years hence gave rise to an enormous bicycle craze that allowed the League of American Wheelmen to become a powerful lobby for improved roads.

Soon the bicycle became a way of life for Bill Metzger. He joined Stanley B. Huber as junior partner in a new enterprise, Huber & Metzger, bicycle dealers. Huber had come to Detroit from Louisville, where he had been one of the most successful bicycle dealers of the region. The jury still was out regarding whether the bicycle could ever replace the horse as personal transportation; however, for Bill Metzger and his growing faith in his own sales ability, it held nothing but promise. The firm officially opened its doors on February 15, 1891, at 13 Grand River Avenue in Detroit.[2-3] Apparently, Metzger initially worked in the furniture store by day and in his bicycle shop by night. However, that relationship was severed that same year, and Metzger began to devote himself full time to his beloved bicycles. Models on display were the famous Columbia, the King, and the Queen, each a product of the Pope Manufacturing Company. (Pope also would become a short-lived power within the auto industry before the end of the century.) The Huber & Metzger bicycle shop grew into one of the largest in the country, influential enough to send bicycle wheels directly to England to be fitted with pneumatic tires because facilities for making such tires had not yet been built in the United States.

Ever on the lookout to improve business, Metzger added the line of Remington typewriters to his store's trade because the latter had been growing rapidly in popularity. He traveled around the city, giving typing demonstrations. With great speed and facility, Metzger would type the words, "Now is the time for all good men and true to come to the aid of their party." Never once did he indicate that those were the only words he could type without a struggle.[2-4]

Metzger had a keen sense for new products and how he could use them to his sales advantage. This would become obvious in his next two ventures.

Metzger in England: Sees the World's First Auto Show

In 1895, Metzger, now only 27 years old, abruptly sold his holdings in the Huber & Metzger bicycle business. The firm must have been doing quite well. Metzger later related that during the spring of 1895, he had read in

William Metzger's first bicycle store, as it appeared in 1891, was located on Grand River Avenue in Detroit. Henry Ford sent this copy to Metzger after finding it in his attic and signed it. (Courtesy of D.A.C. News[2-2]*)*

foreign bicycle journals about an automobile show to be held in London, England, during the fall of 1895. Metzger's interest was piqued to the point that he traveled across an ocean in August ultimately to see what this newfangled machine was all about, most of the hype of that time being about Daimler or Benz automobiles.

Metzger refers to the London automobile show, which was held in October, as being the "first exhibition of automobiles," meaning the first ever. He might have been correct. Unfortunately, the show was not open to the general public (probably assuming the public could not afford such products anyway) and almost invalidated Metzger's trip in the process.

As Metzger wrote years later,[2-5]

> Had letters of introduction to Brown Brothers, then small bankers, but found they could not get for me an introduction to Sir David Solomon who was giving a private exhibition of horseless carriages at Turnbridge Wells race track, a sod track. The day before the exhibition I went to Sir David's office and told him I had come from the States to see the show. He said, "My boy, if you came that far, I will take you the remainder of the way," and had me go with him in his private car the next day to Turnbridge Wells.

On display at Turnbridge Wells were the Daimler, the Benz, the Panhard-Levassor from France, and the Aster and Thornycroft Steamer of English manufacture.

The irony of the Turnbridge Wells exhibition was that it was being held in England, the most notoriously anti-self-powered transportation country in Europe. Because of several accidents with steam coaches in the 1830s, England had passed the Red Flag Act in 1865. This act stated that road locomotives could not exceed two miles per hour in towns or four miles per hour on the highway. Furthermore, each vehicle had to be preceded by a man walking 60 yards in front of the vehicle while waving a red flag by day or a red lantern at night. Metzger mentions,[2-6]

> I actually saw them drag a few steam cars, a gas car, and some DeDion tricycles through the streets of London with horses and a man walking ahead with a red flag.

The Red Flag legislation would be terminated in November 1896, following what probably was Europe's first public auto show in London late that spring. The Imperial Institute of London sponsored this show. Among its attendees was the Prince of Wales, who later would become King Edward VII.

Metzger later went to France and Germany, visiting the factories of both Daimler and Benz. They were the leading auto producers in the world in 1895—Daimler with 24 to his credit, and Benz with an outstanding 135.

The visits impressed the young Metzger, and his intent on his return home was to convince other bicycle manufacturers and distributors to take an interest in the automobile. He arrived in the United States a few days after the historic Thanksgiving Day *Chicago Times-Herald* race, won by the Duryeas. On his return trip to Michigan, he stopped to visit the Duryeas in Springfield, Massachusetts, and took a ride in their car. Metzger later recalled,[2-7]

> Made up my mind then that the light American car had it all over the heavy foreign car such as the Benz and Daimler.

He went on to add,

> Thought the business would start in this country in 1897 and that it would start with commercial wagons. Was wrong in this.

Metzger's absence from the retail bicycle business was short-lived. That same year, 1895, he opened a new store on Woodward Avenue in Detroit. He would own that store for the next 25 years, despite the many other ventures of much greater importance with which he would become associated during that time span.

Metzger's interest in bicycles also brought him to the attention of the pioneering work that the Wright brothers were doing in their bicycle shop to build a flying machine. In fact, Metzger is alleged to have taken time from his many business ventures to journey to Kitty Hawk, North Carolina, where he witnessed the Wrights' first flights that took place on December 17, 1903. There is no evidence to indicate that he was on the scene; however, because he was quite familiar with other "wheelmen" and bicycle dealers in the Midwest, Metzger could easily have witnessed the Wright brothers' flights in Dayton.

Either because of the Wright brothers or some other source, Metzger developed a lifelong passion for aviation and indulged himself in various business ventures involving the aircraft industry in his later years.

Metzger Opens the First Auto Dealership in Michigan and Perhaps the Nation

Past records are not clear, but it seems that Metzger stocked several electric cars in the fall of 1898 in the old Biddle House at 254 Jefferson Avenue between Randolph and Brush streets. (The site later would become the local headquarters for the old Fisk Rubber Company.) His first sale was that of a Waverly Electric, which was purchased by Mr. Newton Annis, a well-known and respected furrier in downtown Detroit. Metzger considered this to be the first retail sale of an automobile in Detroit.[2-8]

The Waverly, built in Indianapolis by the Indiana Bicycle Company in conjunction with the American Vehicle Company of Chicago, had just been introduced.[2-9] Its management no doubt welcomed having someone with the reputation of Metzger offering its product for sale in Detroit, especially with such excellent results: twenty sold for the year, six of which went to wealthy ladies in the Detroit area.[2-10]

The following year, 1899, Metzger expanded his offerings by becoming the agent for several steam car companies, one of them reputed to be the Mobile Company of Tarrytown, New York, although the latter was not in production until 1900. The Mobile Company had acquired the rights from the Stanley brothers to produce a vehicle based on the first Stanley steamer.

By the fall of 1899, Metzger had added gasoline-powered automobiles such as the Winton and the Haynes to his sales fleet. He also took on the task of retailing cars built by Oldsmobile. The company had been incorporated on

Bill Metzger is shown here in his prime. (Courtesy of the Detroit Public Library, National Automotive History Collection)

May 8, 1899, and by July had broken ground for a production plant on the Detroit River at 1330 Jefferson Avenue, not too distant from the Metzger address. Just before that, in June, Metzger sold the first Oldsmobile produced by the new company. It had been built in Lansing and went to a druggist named W.A. Dohany. In a letter to David Beecroft, Metzger noted that he had substantiated the date of this first Oldsmobile sale with Mr. Olds himself and had later checked with Mr. Dohany.[2-11] Metzger also mentioned that the vehicle was a curved dash Oldsmobile. If so, it probably was the earliest version of that now famous automobile.

The First Detroit Auto Show

Before the year ended, Metzger initiated another startling innovation: the auto show. This show was held at Detroit's Lightguard Armory under the auspices of a label invented by Metzger called the "Tri-State Sportsman's & Automobile Association." In other words, to broaden the appeal of the show, sporting goods were added as well. What made the event noteworthy from a historical standpoint is that the only other auto show to have been staged in the United States prior to this time was one in Boston in 1898. Detroit tied with Philadelphia (only four cars) for conducting the second such event, both in 1899. The first of the now famed New York Auto Shows did not take place until November 3, 1900, at the Madison Square Garden. It was a huge event involving 40 car companies and 300 types of vehicles. Metzger was rumored to have had a hand in planning that event also. Chicago likewise held its inaugural auto show in 1900, at which 80 cars were exhibited. Metzger's Detroit Auto Show would be held annually until 1908, when it gave way to a more significant effort adopted by the newly formed Detroit Auto Dealers Association.[2-12]

Due to his efforts in promoting a Detroit Auto Show and his unbridled enthusiasm for the automobile, Metzger found that his dealership was growing well beyond both his expectations and his ability to furnish space for the passenger cars he exhibited. A new two-story building was constructed on the corner of Brush Street and Jefferson Avenue, and it was occupied during August 1901. The expanded quarters enabled Metzger to become the agent for several new brands: the Baker and Columbia electric cars, the Toledo (both steam and gasoline), and the Geneva steam car.

Metzger Promotes Henry Ford's First Racing Venue

Ever the consummate promoter, Metzger, Daniel Campeau of the Detroit Driving Club, and Charles Shanks, sales manager for the Winton automobile, put their heads together and came up with one of the better-known race events in auto history. It became famous simply because its outcome had such a profound effect on the career of Henry Ford.

The race card was scheduled for October 10, 1901, at the Grosse Pointe (horse racing) track, and was more than a race event. It became the flash point for a gala all-day affair involving the city of Detroit. Excitement came from several quarters. The most well known of the competitors to enter the race was Alexander Winton, who had spent the entire year touring the country with his special racing vehicles, usually in pursuit of the mile speed record for automobiles but primarily to promote the cars of his own manufacture. To have him appear in Detroit was big news. However, rumor quickly spread that Detroit's own Henry Ford, a struggling auto inventor of still dubious success, had filed the preceding day to race against the Winton forces in the premier 25-mile event.

That alone seemed to have perked additional interest. In fact, Judge James Phelan earlier had announced that the Detroit Recorder's Court would be adjourned on the afternoon of October 10 at the request of attorneys and others. The streets from downtown Detroit to Grosse Pointe also became the scene of a parade of more than 100 automobiles led by electrics. They could very well have represented the majority of automobiles owned in Detroit at the time, and many of the vehicles probably had been purchased through Bill Metzger. Special streetcars were run from downtown Detroit to the race track every half hour.

There were races for electric cars, steamers, and gasoline cars, but interest began to wane as the 8,000 spectators waited for the "big race," especially when the slow-moving electrics took over the track. To rekindle excitement, Metzger persuaded Winton to make a special three-mile run against time. It proved to be just the tonic needed. Winton set a new existing record for the mile at 1:12.4 on the second lap. (Ironically, at about that same instant, Henry Fournier in New York came in with even a lower time of 1:06.8.) By now, the day was so far gone that Metzger was forced to announce that the feature 25-mile race would be reduced to 10 miles. No matter—it proved to be an enormously exciting contest in which Henry Ford ultimately was victorious over Winton by an amazing three-quarters of a mile, much to the delight of the crowd who had not anticipated a win by a local lad.[2-13] The feat no doubt led to Henry Ford acquiring backers for a second auto company, the Henry Ford Company, incorporated the following month. (His first company had failed, as would this second venture.)

Metzger Adds More Floors to His Dealership

By 1903, Metzger's business had grown to such proportions that he had four more stories added to the existing structure. It now was the largest such establishment in the region. The showroom with turntable occupied the main floor. On it could be found the most recent examples of automobiles as an Autocar, Columbia, Pope-Waverly, and Pope-Toledo, or Packard and Cadillac. Replacement parts sales also were made on this floor. The second floor contained the company offices, as well as the electric vehicle-charging department. The third floor was devoted to the repair of gasoline automobiles; the fourth floor was devoted to the repair of electric cars. The electric and gasoline car repair stations were kept apart because Metzger believed that a separate expertise was required for the maintenance of each vehicle type. The top two floors were used for storage. A large, automobile-sized elevator transported vehicles or whatever else was necessary between floors. The works were open 24 hours a day to service customers in need.[2-14]

The entire facility, sprouting from an awkward and chancy storefront operation in 1898 to the enormous operation it had become in 1903, mirrored the sudden growth of the automobile industry in the Detroit area during that same period. Quickly realizing this potential, Metzger wasted no time in staking his business life on further new endeavors involving the automobile. He rapidly was acquiring the reputation and the funds to make things happen.

Bringing the Northern Automobile into the World

Reminiscing in a letter to David Beecroft in 1915, Metzger recalled that in 1900 he organized the Northern Manufacturing Company "which was owned with Mr. Barbour, Mr. Gunderson, and myself." Their goal? To build a reliable automobile that would allow them to compete with early makes coming into Detroit for sale.

At this point in time, 1900, we would have difficulty finding a successful manufacturer of automobiles in the Detroit region. In fact, there was not the slightest hint that Detroit later would become the auto capital of the world. Henry Ford had formed his first of three automobile companies, the Detroit Automobile Company, in 1899; however, it had sunk from sight by late 1900 without having offered a vehicle for sale. The Olds Motor Works had moved into a new plant in Detroit to commence automobile production but as yet had built no more than about a half-dozen cars. Approximately 30 miles north in Oxford, Michigan, C.P. Malcolm & Co. was producing three different models of gasoline-powered cars, but the company quit production after only one year. That was it!

By deciding to produce their own vehicle, Messrs. Metzger, Barbour, and Gunderson were launching into relatively uncharted waters. Unlike a few years hence, when Detroit automobile factories would spring up hither and yon across the city and its environs, similar to weeds in an empty lot, little information was available in the field to provide direction for establishing an automobile business. The three men had one thing going for them, however: solid financial backing. This was a rare commodity, one that prevented Charles Brady King from forming his own company after he developed the first automobile to be seen on the streets of Detroit in 1896.

The Barbour name alone would have appeared to be a guarantor of success. The family had acquired control over two of the largest stove companies in Detroit—the Michigan Stove Company and the Detroit Stove Works—at a time when the city was the undisputed capital of the stove industry in the United States. William Barbour was president of the Detroit Stove Works, while George B. Gunderson was secretary-treasurer of the Michigan Stove Company. These men were among the financial elite of Detroit.

What the new company needed was a pilot model of an automobile to put into production. For this step, they went to Oldsmobile. According to Metzger,[2-15]

> We took Mr. Maxwell, who was then working for the Oldsmobile Company, also Mr. George Trout, who was also with the Oldsmobile Company, and had them each build a model car for us in the fall of 1899 and early part of 1900. We selected the Maxwell design as the pilot for the Northern.

Actually, Jonathan Maxwell was well acquainted with William Barbour. Often while taking an Oldsmobile out of the Detroit factory for a test run, Maxwell would stop at the Detroit Stove Works next door to the Olds plant and invite Barbour for a drive.

The First Northern Described

Maxwell's entry was a single-cylinder, five horsepower runabout on a 68-inch wheelbase that sold for $800. A top was available for another $55, and a leather-covered "dos-a-dos" ("two by two") attachment for an additional $35. Power was transmitted directly to the rear axle via a detachable chain. There were two speeds: forward and reverse. The body could be detached with little effort. Some of the innovations on the car were three-point suspension of the engine, running boards, and a transmission that was integrated with the engine. Steering was done via a right-side-mounted tiller. The engine was water cooled via a circulating pump driven off the camshaft that forced the water through the radiator.[2-16] In some ways, the Northern resembled the curved dash Oldsmobile. It was understandable, considering from whence Maxwell came.

With the pilot model in hand, a building was leased at Helen and Champlaign streets, not too far from the Oldsmobile factory. Finally, with all the building blocks in place, the Northern Manufacturing Company was incorporated, the date of incorporation being February 19, 1902. According to the first annual report of the company, filed with the state of Michigan, the president of Northern was W.T. Barbour; vice president, George Barbour, Jr.; and secretary and treasurer, George B. Gunderson. Capital stock was set at $50,000, of which $38,000 already had been paid in. These three plus Jonathan Maxwell formed the board of directors. Conspicuously absent was any mention of William Metzger even among the list of stockholders, although it seemed to be common knowledge that he was a major factor in the formation of the company. However, W.T. Barbour was listed as a trustee for 875 shares, one of the largest blocs issued, and this might be where the supposed omission resided.

Sales of the Northern were more than adequate: 200 in 1902, and 750 in 1903. This placed the company about fourth or fifth in the nation.

King Joins Northern, Maxwell Leaves

At some point between 1902 and 1903, Charles Brady King left Oldsmobile for Northern. Meanwhile, on July 3, 1903, Jonathan Maxwell signed a contract with Ben Briscoe to develop a pilot model of a new vehicle. Around this

The E-M-F Company

A 1905 Northern runabout, shown at a contemporary auto show. The car sold for $650 in 1905.

car Briscoe planned to form a new company, which he did, taking in Maxwell as his partner. The new venture was called the Maxwell-Briscoe Motor Company, one that was well regarded during the first decade of the twentieth century. The company even broke into the top echelon of the industry in sales.

Speculation was that Maxwell and King could not get along with each other, and when Briscoe made his offer to Maxwell, especially to produce a car bearing Maxwell's name, the latter jumped at that opportunity.

In his July 7, 1915, letter [2-17] to David Beecroft, Metzger wrote,

> Mr. Maxwell was with the Northern Company for a number of years. When he left, he went with Mr. Briscoe. We replaced him with Charles King, who then became the designer for the larger Northern cars.

The exact date that Maxwell completely severed his ties with Northern can only be speculated on, inasmuch as he remained listed as a director and owner of 250 shares of Northern stock in the annual report for the company on January 1, 1904.

Hail, Yale

What became of George Trout's losing design, the one that lost to Maxwell in the competition for the first Northern prototype? It was anything but forgotten. In Metzger's own words,[2-17]

> Mr. Trout built a little two-cylinder car for the Northern Company and when we rejected this [for the Maxwell entry], I purchased Mr. Gunderson's and Mr. Barbour's interest in this model and turned the model over to the Yale Manufacturing Company of Toledo, in which I was stockholder. This car was developed later into the Yale car.

It appears that Metzger may have been in slight error on the point of manufacture. The Kirk Manufacturing Company built the Yale car that ultimately appeared on the market in 1902. Late in 1903, Kirk merged with two other Toledo firms, the Snell Cycle Fittings Company and the Toledo Manufacturing Company, to form the Consolidated Manufacturing Company. All three were involved in the bicycle business, which explains the ease with which Metzger was able to go to Toledo and convince them of the virtues of building the Trout-designed vehicle. Ironically, Consolidated was capitalized at $600,000, six times that of Northern. The car it built continued to be called the Yale.

The Yale turned out to be a creditable vehicle, advertised as "The Car with the Doubt and the Jar Left Out." It held its own, although it was a bit pricey for a runabout at $1,500. It continued to hold its popularity in Ohio through 1905, even after adding a five-passenger touring car that listed for $2,500. Early in 1906, Consolidated suddenly announced that heretofore it would focus production on only Yale and Snell bicycles. No explanation was offered. Four months later, the company went bankrupt.[2-18]

Without Metzger, There Would Be No Cadillac

As if his bicycle business and his automobile dealership, as well as his help in forming the Northern and Yale car companies, were not enough for 1902, in that same year Metzger accepted the challenge of bringing success to the infant Cadillac Automobile Company.

Cadillac was reorganized out of the Henry Ford Company on August 22, 1902. Henry Ford had long since departed from this, the failure of his second motor car company. It had been formed in November of the preceding year by a group of wealthy Detroiters headed by William Murphy, the son of a millionaire Michigan lumber merchant. The group had hoped to capitalize on Henry Ford's sudden claim to fame from winning the Grosse Pointe race on October 10. Ford's task was to engineer a successful prototype vehicle that could be placed into production.

Unfortunately, Henry Ford had been severely bitten by the racing bug. Instead of focusing on developing a passenger car, he continued to devote most of his attention to racing vehicles. Murphy warned him to shift gears, but Ford stubbornly resisted to a point where Murphy soon decided that Ford would have to go if the backers were ever to realize any gain from their investment.[2-19]

Evidently, Murphy finally got under Ford's skin to the point that Ford resigned on March 10, 1902. Whether the resignation was freely given or forced on him by the company has long been open to speculation.

Thus, the Henry Ford Company entered the spring of 1902 with neither the man for whom the company was named nor a vehicle to produce. Oliver E. Barthel, whom Ford had brought into the engineering department earlier, was given the task of coming up with a pilot model, which he did: a two-cylinder, four-passenger runabout. Apparently, the board was skeptical of the worth of this vehicle, for they asked Henry M. Leland to evaluate it. (There also was some question about whether Leland was asked to evaluate the entire company, as to whether it should remain in business.)[2-20] Leland proposed the unexpected: that the Henry Ford Company build a new prototype using his one-cylinder engine, which had just been refused by Oldsmobile but which Leland was certain would bring them great success. The board agreed but also decided that it was necessary to reorganize the company. The name chosen for the new organization was the Cadillac Automobile Company, in honor of the founding father of Detroit.

Events now began to move rapidly. When engineering had difficulty putting together the prototype, Leland stepped in and assigned the task to his own engineer, Alanson P. Brush.

On October 8, 1902, the company made another enormously significant addition to it future fortunes: William Metzger signed a contract with Cadillac as its first sales manager.[2-21]

The first Cadillac was ready by October 17, 1902, and was taken for a run by Brush and Leland, although this may not have been the entire story. If Metzger is to be believed (and there is no reason to disbelieve him), he may have had a heavy hand in the Cadillac that emerged at that time. In his October 22, 1924 letter to David Beecroft, Metzger also wrote,[2-21]

> The success of Cadillac cars shown for first time to public October 24, 1902 at Grosse Pointe Race [Automobile Race] which I ran were greatly due to a detachable demi-tonneau which I copied from a Winton 1902 model of which I sold 24 at retail during the year of 1902.

Metzger went after the selling of Cadillacs similar to a man possessed. Within weeks, he was running ads in trade journals, warning prospective dealers that they must act soon because "the Cadillac agency is bound to be snapped up quickly." He was unashamed to use strange but newsworthy stunts to promote the Cadillac. One such stunt was having Alanson Brush drive a Cadillac, which was equipped with tire chains, up and down the steep steps of the Wayne County Court House on Cadillac Square in Detroit, while several thousand spectators cheered.

Metzger virtually guaranteed the early success of Cadillac by what he accomplished at the 1903 New York Auto Show. Even from this distance in time, it is something to be admired. According to Henry Leland in his book *Masters of Precision,*[2-22] two further models were completed at the Leland & Faulconer Manufacturing Company in time to be displayed by Metzger with the original prototype at the New York Auto Show during January 1903. Metzger is reputed to have sold 2,286 Cadillacs (Leland remembered it being "over 1,000") during the show week, each involving a deposit of at least $10. He even had the audacity to announce before the week ended that he no longer could accept orders because the Cadillac was now sold out for the year. As events transpired, it appears that 2,497 new Cadillacs actually were completed through December 1903, which is not too far off the Metzger mark.[2-23] Considering that the full production of Cadillacs did not actually begin at the old Ford factory on Cass and Amsterdam avenues until March 1903, the number produced for the year is as startling as the fact that Metzger had pre-sold all of them in a matter of a few days early in January. (This factory, incidentally, was only a few blocks directly south of the future General Motors Building on West Grand Boulevard.) You can imagine the elation felt by Cadillac management when it discovered that its company, in the first year of existence, had out-produced every other automaker in the United States except Oldsmobile. In fact, two of every five cars built in 1903 were Cadillacs!

As exhibited at the show, the new Cadillac was an attractive runabout carrying a single horizontal cylinder engine advertised at 6.5 horsepower. It had mechanically actuated overhead valves and a copper water jacket.

Immediately after the show, Metzger ran an advertisement in leading newspapers and magazines across the country, in which he used quotes supposedly made by dealers that he had signed up as Cadillac agents during show week.

Some of the quotes were:

> Don't See How They Can Do It for the Money

or

> The Cadillac's a Cinch

The ad included the names and addresses of 21 different now-Cadillac dealers.[2-24]

Metzger obviously knew his business. In fact, he contracted with N.W. Ayer & Son, one of the most highly regarded advertising agencies in the country, to handle Cadillac ads. Early attempts made by the agency were to feature the Cadillac in some of its more unique stunts, such as driving up and down the steps outside the U.S. Capitol in Washington, D.C., or to show the Cadillac runabout carrying 16 men, driving up steep Shelby Hill in Detroit.[2-25] The ad for the Capitol stair-climbing incident read:

> A man drove a Cadillac up the steps of the Capitol at Washington. He paid for his fun (he was arrested), but it was worth the money to know the power of the Cadillac.

Such stunts also became featured attractions in the newspapers. For example, the February 9, 1903 edition of *The Detroit Free Press* reported that a single-cylinder Cadillac had pulled a five-ton truck filled with railroad iron up a four-percent grade after a two-cylinder car had failed.

Metzger may have succeeded in turning Cadillac into a money-generating machine. In doing so, he certainly did well for himself. According to the terms of a contract offered by the Cadillac Automobile Company to Metzger in letter form and dated December 23, 1903,

> You are to give your entire time and attention to marketing the product of the Company for which services the Company is to pay you a salary of $1,000.00 per month for the period above stated (10-8-03 to 10-8-04). In case the net amount received by the Cadillac Automobile Company from

the sale of its product during said period exceeds $600,000.00, you are, in addition to your salary, to be paid, at the end of your year's service, a commission of 1½% on such excess provided said excess does not amount to more than $900,000.00.

The contract then stated that Metzger would receive a one-percent commission on sales that went beyond the latter amount. Apparently, Metzger was hired on a one-year basis and offered a new contract each year. Based on Cadillac sales during this period, it is conceivable that he realized a hefty extra income for his work, probably in the neighborhood of $30,000 to $40,000 in 1904, for example. Balanced against the sales success he brought to Cadillac during its early years, we doubt that management was prone to quibble over the money paid to him. Proof of its confidence in Metzger's abilities can be found in the fact that he remained the sales manager of Cadillac until the spring of 1908, when he resigned to join Everitt in forming EMF.

The jump-start that Metzger gave Cadillac sales almost came to a halt on April 13, 1904, when most of the factory buildings on Amsterdam and Cass avenues burned down. Work on recreating them into a seven-building, 275-square-foot complex was instituted as quickly as possible. Nevertheless, production was nil for 45 days and then resumed on a tentative basis. The upper two stories of the surviving general office building on Cass Avenue were installed with finishing machines. Most of the new car assembly took place within a large warehouse across from the general offices. The bodies were built at the Peninsular Mill Machine Screw Company and Detroit Curling Club, then were taken to the top two floors of Metzger's dealership, where they were finished. Management very quickly had production up to 22 vehicles per day, and Metzger announced that the build rate would go up to 40 vehicles per day by June 1.[2-26] Nevertheless, deposits for approximately 1,500 cars had to be returned because of the inability of the company to fulfill the orders.

Despite its quick return to full production, Cadillac management was dissatisfied with what was taking place on the factory floor. The manufacturing process kept breaking down, and suppliers were either too often late or providing a shoddy product. The following year, it decided to invite Henry Leland about taking control of production. William Murphy told Leland that "either you boys will come and run the factory for us, or we will go out of business."

Inasmuch as Leland was Cadillac's major supplier (it provided all Cadillac engines and transmissions), the loss to both enterprises would be painful. Leland finally agreed in the fall of 1904, although he later would say that he never intended to get into the automobile business because there was too much trouble in it.

Then, on October 1, 1905, Cadillac announced that the company was being reorganized. Cadillac would merge with the Leland & Faulconer Manufacturing Company to become the Cadillac Motor Car Company with a capitalization of $1.5 million, based on the assessment that the original Cadillac company now was worth $1 million and Leland & Faulconer was worth $0.5 million. Henry Leland would become general manager and Wilfred Leland assistant treasurer.[2-27] Most interesting was the list of stockholders: William Murphy, with 6,600 shares was the largest stockholder, followed by Clarence Black, Lem Bowen, and Albert White with 5,500 shares each, while Metzger held 3,000 shares, which was 500 more than what Henry Leland received for merging his company with Cadillac. Clarence Black became president.[2-28]

By the close of 1905, however, Metzger already had begun looking toward new enterprises and in selling off his large sales agency. In replying to a letter from Henry Joy, Metzger wrote,[2-29]

> The Cadillac Motor Car Company has purchased from me the retail business, with the exception of the accessory portion, and certain automobiles other than Cadillacs. This Company has sub-leased the building, except a small section necessary for carrying out the accessory business.

Thus, the splendid six-story building that symbolized Metzger's commanding presence as the largest (as well as first) auto dealer in the Detroit area passed on to Cadillac.

Northern and Metzger in the News Again

In 1906, a substantial change occurred within the Northern Manufacturing Company, bringing the Metzger name back to his relationship with this company. The company was reorganized as the Northern Motor Car Company with a capitalization of $500,000, ten times that of the original. Of this amount, $160,500 had been paid in cash, so one would think that sufficient working funds existed to make the organization successful. William Barbour continued as president, with George and Victor Gunderson filling the slots of vice president and secretary/treasurer, respectively. Charles Brady King replaced Jonathan Maxwell on the board, in keeping with his title of chief engineer. Maxwell had long since left an active position with Northern.

What probably caused the reorganization was the opening of a new Northern plant in Port Huron, Michigan, approximately 50 miles north of Detroit. It was a brick two-story building that covered 65,000 feet of floor space and employed 225 men. Here Northern intended to build its two-cylinder engines, while a new four-cylinder engine, designed by Charles King, would be assembled in Detroit. The annual report of the company to the state of Michigan for 1906 listed 45 new stockholders from the Port Huron area, most of whom had invested in at least 100 shares of Northern stock.

Other changes were significant on this list. Charles King now held 1,798 shares of Northern stock, but his wife had added 4,000 shares to the family holdings. William Metzger's name looms large as well, appearing with those of the Barbours and Gundersons in an amount similar to their holdings, or 5,000 shares. As an owner of more than 12 percent of the outstanding stock, Metzger was in the position of having a strong voice in the future affairs of the reorganized company.

Newspaper coverage of the Northern products could be humorous as well as enlightening. For example, in reading the Financial Section of *The Detroit Free Press* dated January 20, 1907, the reader is not quite certain how to take the following:

> This car [the two-cylinder Northern] has the almost unequaled record of being presented for the fourth season without any change in the construction or design.

Later in the article, the writer reports that

> ...the Silent Northern car has been steadily forging ahead in public favor, and tho' it is the highest-priced two-cylinder car

in the world, the sales for 1906 exceeded those of 1905 by almost 33½ percent...

(The two-cylinder limousine sold for $2,800, or almost three times that of its two-cylinder brother, the runabout.) The use of the name "Silent Northern" was earned by the huge, double muffler system, four and one-half inches long and six inches in diameter. That and other efforts such as an unusually large flywheel that was 24 inches in diameter produced a quiet, quality ride.

Describing the 1906 Northern

The car that made most of the news for Northern in 1906 was Charles King's four-cylinder Model K with a horsepower rating of 30. It featured a unique air compressor that powered the clutch, the brakes, and the fuel feed to the carburetor. The compressor was driven by a crank off the camshaft and activated a piston with bore and stroke of three times three inches. It supplied 60 pounds of pressure to engage the clutch, and forced fuel from the tank to the carburetor under a pressure of two pounds per square inch. Air pressure activated a two-inch diameter piston to engage the rear brakes. The beauty of air pressure was that it neither froze nor wore out when used repeatedly. Although no complaints were registered about any faulty operation involving air pressure on a Model K throughout 1906, the system was discontinued by the 1908 model year simply because Northern sales outlets complained that few people wanted to buy a vehicle that depended on new, untried technology.[2-30] At an advertised price of $4,200, one could easily expect that there might be only a few takers anyway. This was a rather high amount from a company that still offered a runabout at the low end for only $650. It is another example of a company overreaching its goals simply in an attempt to create a high profit per unit sold.

Northern management scheduled the build of 500 two-cylinder cars (Models B and C) and 50 four-cylinder cars (Model L), which would have been more than produced by such well-known old car companies as Moon, Marmon, Stearns, and Willys-Overland. However, this ambitious plan was never realized. Northern simply was not in good shape. Apparently, the Port Huron expansion became another case of too much with too little. In its annual report to the state of Michigan for the year ending 1906, Northern wrote that

Enjoying a Sunday drive in a 1906 Northern touring car. (Courtesy of the Detroit Public Library, National Automotive History Collection)

cash on hand, which included money in the bank, was only $384.68, despite the fact that the amount of capital stock paid in at that point since the reorganization was $160,500.

Something had to done soon if Northern were to survive, and that something would come from one of its leading stockholders, William Metzger. Northern now became the second leg of the three-legged stool that eventually resulted in the formation of the Everitt-Metzger-Flanders Company.

Chapter Three

The Merry Master of Mass Production— Walter E. Flanders: His Early Years

A half century after Walter Flanders had worked for Henry Ford, Charles E. Sorenson, long a key man in building the Ford empire, wrote that Flanders was a "forceful, boisterous man" who was popular with both the directors and the men in the shop, and that "his rearrangement of machines headed us toward mass production."

Sorenson further commented[3-1]:

> Without the genius of Walter Flanders in arranging production machinery and in cutting supply and inventory costs, the way would not have been paved for economical production of the Model T, and the moving assembly line, upon which American mass production depends, would have been long delayed.

Without a doubt, Walter Flanders was an imposing figure, as anyone would be who was six-feet three-inches tall and weighed approximately 250 pounds, mostly muscle. Sorenson described Flanders as "a large, heavy man with a great head of curly hair and a voice that could be heard in a drop-forge plant."[3-2] Another source called Flanders, "bluff, pugnacious, two-fisted, [with] a bellowing voice that could hammer through the clatter and din of a shop."[3-3] Still another painted the picture of a "big, burly, rough-mannered man, with a large head set off by a shock of bristling hair, and an imperious manner."[3-4]

Walter E. Flanders at about age 40. The photograph is dedicated to his close friend, Alfred Owen Dunk. (Courtesy of the Detroit Public Library, National Automotive History Collection)

Walter Flanders was born in 1871 in Rutland, Vermont. He was the son of a poor country doctor, and he quit school at the age of 15 to apprentice in one of the Singer Manufacturing Company plants. From Singer, he gravitated to the Landis Tool Company. By the time he was 35 years old, Flanders had become an expert on machinery, their installation, and their maintenance. Working out of Cleveland, he represented three different machinery manufacturers: Potter & Johnson; Manning, Maxwell & Moore; and the Landis Tool Company. More importantly, whenever anyone purchased machines from Flanders, he was on hand when the machinery arrived, supervised the setup, demonstrated how the machinery worked, and trained the men who would operate the machinery.

On one of his visits to Detroit, Flanders toured the Leland & Faulconer Manufacturing Company and told Henry Leland that his machines were too crude and too slow in their operation. Flanders claimed that he could provide Leland with machines that would do ten times the work in one-tenth the time. Leland took up Flanders on his suggestion, whereupon Flanders returned to Cleveland, invented the machines to do the job, and fulfilled the contract.[3-5]

The manner in which Walter Flanders met Henry Ford and eventually became Ford's production head has been the subject of various tales that Flanders himself apparently never put to rest.

Flanders Advises Henry Ford

The most common account is that Flanders stopped at Ford's Piquette plant on one of his periodic swings as a machine tool salesman through the Detroit area in 1905. At the time, Henry Ford needed crankshafts and accepted Flanders' challenge that he could produce 1,000 of them according to a specified schedule set up by Ford. There also was the carrot that more work would be directed Flanders' way if he came through as promised. Flanders fulfilled his end of the contract, and the fact that he did so with such confidence remained in Henry Ford's mind.

Barney Everitt, Flanders' close friend, said that Henry Ford one day complained to Flanders that the machines he had purchased from him were not producing at the rate he was led to expect. Flanders thereupon went into the Ford factory, took off his coat, and demonstrated that not only would the machines produce per the sales agreement, but they could exceed the rate if handled properly.[3-6] Flanders not only sold machines to Ford but helped train his line workers and told Ford where to locate the machines for an optimum flow of materials.

In 1905, the leading automobile manufacturer in the United States was Oldsmobile, followed by Buick and the Thomas B. Jeffery Company, maker of the Rambler. However, Henry Ford had aspirations that far exceeded any of these companies—to become the first in the industry to produce cars annually in excess of five figures.

The opportunity to achieve this goal came about with the death of John S. Gray on July 6, 1906. Gray had been company president since Ford Motor Company was founded. Henry Ford now became the company's new president. This gave him free rein to realize his own dream of producing a low-priced car, while at the same time increasing production to a point well beyond anything the competition had thus far envisioned. The car that would fulfill this promise would be the Model N, which was priced at $500—a third less than that of any of the previous Fords.

Henry Ford realized that the only way he could reach such a high level of production was by replacing men with machines wherever possible. Without bigger and better machines, mass production was an unattainable goal, as would be the successful manufacture of replacement parts that would mirror image the original equipment part. The biggest hurdle that Ford had to overcome was the lack of plant space. The Piquette plant was only so large. Increasing the use of machines and building a greater number of cars placed floor space at a premium. Unfortunately, no one within Ford seemed able to come up with an orderly way to position machines to enable a satisfactory progression of materials. Confusion reigned.

Flanders Joins Ford Motor Company

At this point, Henry Ford remembered Walter Flanders, the most expert man on machinery he had ever met. Flanders might be the answer to his problems. Sorenson wrote that James Couzens, Childe Harold Wills, and both Dodge brothers recommended Flanders but "Ford was worried. He was aware of Flanders' ability, yet feared him. There was a streak of jealousy here."[3-7]

Nevertheless, Flanders was offered the position of works manager for the Ford facilities. He accepted and began work on August 15, 1906, at the princely salary of $7,500 per year, with the proviso that either party could terminate the contract after giving a three-month notice. Flanders also was permitted to retain his own business and continue to act as the manufacturer's representative for several machine tool companies.

Flanders had extracted a very hard bargain. He demanded complete control over manufacturing operations. His was an unprecedented request from someone coming into the company from the outside, but his request was granted. He also brought with him his erstwhile Cleveland partner, Thomas S. Walburn. The two men stopped production of the new 1906 Model N completely for a time as Flanders realigned machinery and Walburn trained the operators to obtain maximum output without loss of quality.[3-8]

According to Barney Everitt, Flanders soon increased production from 20 to 150 cars per day. This was hardly possible because it would result in an annual production of 40,000 units, about double that of the entire industry of the time.[3-9] More likely, Flanders increased production to approximately 35 cars per day, which would have been a remarkable feat in itself.

Flanders Increases Ford Model N Sales

Everitt also related that Flanders was disappointed because Ford could not sell as many of the low-priced 1906 Model N's as he was capable of building. Flanders complained to Everitt that the problem with the Model N was that it resembled a glorified buggy instead of an automobile, especially because it had no running board and only partial fenders. Flanders took it upon himself to contract with Everitt to build a new automobile body with full fenders and running board. He had Everitt place the body on a Model N chassis, which Henry Ford then approved. The upgraded Model N cost an extra $25, which was recouped when Flanders convinced Henry Ford to charge $100 more for every Model N that carried the new body. (He also proved to Henry Ford that Ford was losing up to $12 on each current Model N that came off the line.) Within 60 days, according to Everitt, demand for the new Model N was exceeding supply.[3-10]

It is quite conceivable that this latter recollection by Everitt has merit. Ford did introduce a four-cylinder Model R and Model S in February 1907, both of which were more glamorous versions of the buggy-like Model N. They were priced $100 to $150 more than the Model N. Approximately 2,500 Model R's were built that year, which alone would have placed the vehicle among the top five or six cars produced. Thus, Flanders had more to offer the Ford Motor Company than his machinery expertise.

Flanders Introduces Ford to Production Scheduling

Another side to this talented man contributed to Ford's good fortune at this time. When Flanders arrived at Ford, he found that a production schedule did not exist. Ford merely built every car that he could. The newfound popularity of the inexpensive Model N caused this practice to change. By late fall of 1906, demand began outstripping supply. This was fine with Walter Flanders as head of production. It provided him with a reason to forecast a 12-month build program and arrange for the purchase of materials and parts on a strictly scheduled basis. Dealers were told the number of cars they would receive to sell each month.

Flanders created a policy memorandum, approved by Henry Ford, which called for the build of 11,500 Model N's and 600 Model K's plus spare parts over the forthcoming 300 working days. The memorandum also included the statement that the manufacturing department would be responsible for all labor management, materials purchasing, and manufacturing for the company. Further, manufacturing could employ men, buy materials, and deal directly with suppliers for its needs. Flanders then hired an expert stock-keeper to keep a running account of all materials and parts on hand. When he had the materials and parts pared down to a 10-day supply, he was told that he had to strictly maintain that number.[3-11]

This ended Ford's helter-skelter, day-by-day scramble to keep the necessary parts and materials on hand. In turn, the purchasing department found that it could negotiate lower prices for a projected, fixed amount of supplies. The suppliers now became the carriers of large inventories instead of Ford. In this sense, Flanders was a very early practitioner of the "just-in-time" formula adopted by most large car companies today. The new practice came as a refreshing revelation to a Ford Motor Company that previously felt compelled to tie up large amounts of money for emergency inventory requirements.[3-12]

Consolidating Assembly at the Ford Piquette Plant

Early in 1907, Flanders and Walburn began removing the better machinery from the Ford Bellevue plant and installing them at Piquette. The plan was to consolidate all Model N manufacturing at Piquette and thereby markedly increase overall production.

The Bellevue plant originally had been leased from the Wilson & Hayes Manufacturing Company to build engines for the Model N. Much of the machinery in the plant had been purchased through Flanders. In fact, the manager of the Bellevue plant, Max F. Wollering, had been hired on Flanders' recommendation. (Wollering had built machine tools and had overseen the manufacture of stationary gasoline engines for the Hoffman Hinge and Foundry Company of Cleveland.) The Bellevue engine plant had been a distinct departure for Ford because Dodge had built all Ford engines to that point, although the latter continued to provide the engines for the more expensive Ford cars.

Moving in and installing the better Bellevue machinery and adding new machines to a vastly enlarged Piquette plant was quite a challenge. All the heavy equipment meant for such tasks as manufacturing engine blocks, cylinders, crankshafts, and crankcases was located on the main floor. Lighter machinery was installed on the second floor, and the third floor was reserved for final assembly. Under Flanders' and Walburn's supervision, all tasks were arranged so that there was a logical progression of the build of an automobile—from raw materials to finished product.[3-13]

Flanders Quits Ford Motor Company

Suddenly, similar to a bolt of eye-piercing lightning out of a slumbering sky, on April 15, 1908 (according to popular accounts), Walter Flanders resigned from the Ford Motor Company. Only six weeks earlier, Henry Ford had announced his plans for a successor to the Model N, namely, the Model T. The manufacture of the Model T had been the long-range goal of the Piquette plant renovation, but the man who had been so responsible for preparing Ford Motor Company for this ultimate landmark in mass production had now left.

Actually, Flanders did no more than leave the front door of the Ford Motor Company and walk one block west to the front door of the Wayne Automobile Company. Sorenson recalled that Flanders had given the Ford Motor Company 10 days notice of his departure. However, as Flanders talked to James Couzens about his future plans, Couzens took out the company checkbook, wrote a check, and handed it to Flanders with the comment,[3-14]

> This will pay you up to today. If you are going to leave, you might as well go right now.

One story has it that Flanders took violent exception to Ford's plans to forsake the highly successful Model N for an entirely new, albeit higher-priced, Model T. Flanders claimed that the new model would never sell. Distraught with Henry Ford's decision, he decided to leave the company.[3-15] This exchange hardly seems likely. The Wayne Automobile Company simply gave Flanders and his two friends Barney Everitt and Bill Metzger an opportunity to run their own car company—an opportunity too good to be ignored.

All such stories aside, there is good reason to believe that Flanders had made known his move to Ford long before April 1908. The March 12, 1908 edition of *The Motor World* carried a brief note that Flanders already had acquired an interest in the Wayne Automobile Company, and that he and Walburn had resigned their positions with the Ford Motor Company. The article then stated that fresh capital had been brought into Wayne, and that "plans are afoot for the production of automobiles 'on a scale heretofore unattempted.'"[3-16] The article also said that Flanders would become general manager of the revitalized Wayne Automobile Company, with Everitt, who had held that position, moving up to president. The problem with this account is that Everitt had officially become president of the Wayne Automobile Company as far back as September 1907.[3-17]

Regardless, it is a wonder that Henry Ford survived Flanders' resignation, for when Flanders left Ford Motor Company, he took with him Max Wollering and Thomas Walburn. All three men had been instrumental in helping Ford set up the Piquette plant for greater production efficiencies—efficiencies that initially would help make the Model T an economically priced vehicle.

The loss of Flanders, Walburn, Wollering, and then Pelletier, Ford's advertising manager, over such a brief period must have affected Ford initially. Fortunately, Henry Ford had built a pool of expert talent that was able to step in and take their places.

Flanders' departure from Ford Motor Company meant that the third and final leg was in place for the formation of the Everitt-Metzger-Flanders Company.

Flanders Takes Over the Wayne Automobile Company Production

At first, getting the Wayne Automobile Company in the black was the reason for the hiring of the Flanders team of production experts. Everitt as president and past general manager of Wayne may have been a whiz at some things; however, building a complete automobile at a cost that would lead to a reasonable profit did not seem to be his particular brand of expertise. Flanders as new general manager, Walburn as factory manager, and Wollering as general superintendent all went to work to correct the situation.

Yet, try as they might, the three men could not make enough headway. With three different versions of the 1908 Wayne Model 30, each selling for approximately $2,500, one would think that the size of the profit made on each car would be substantial. Unfortunately, despite Flanders' best efforts to cut costs, the company was losing $400 to $450 on each car sold. This was a recipe for ruin!

Palms, Book, and the other primary directors urged that the company should build another vehicle on which it had been working to keep the company solvent, a four-cylinder intermediate-sized automobile of 35 to 40 horsepower that would sell for approximately $1,800. Flanders, who was still fresh from the Ford factory, argued against this proposal and disputed the claims that the vehicle could be a commercial success. Flanders told the directors that a new Wayne automobile similar to the one they proposed would have to compete with the likes of Cadillac, Chalmers, or Mitchell, all well-established brands in the marketplace and selling at prices $300 to $400 less than $1,800. In other words, the Wayne vehicle they proposed did not stand a chance against the competition.

In Flanders' opinion, if they expected to make money on a new product, it could come only from heading in the opposite direction. He argued that a smaller touring car selling at approximately $1,250—half the price of the current Wayne 30—would generate much more sales volume and thereby greater revenue even though the profit margin per vehicle would be lower. The car he had in mind would be the size of an intermediate, designed to be as good as a Cadillac or a Chalmers or a Mitchell, but underselling them by several hundred dollars. Flanders claimed that such a vehicle could bring the company a profit, but one that would be based on building and selling a large number of vehicles rather than a few at a high price.[3-18]

Eventually, Flanders' arguments won. What swayed the Wayne Automobile Company directors was his assurance that he actually would be able to produce an automobile complete with magneto and lamps for a selling price of $1,250.

The directors then decided that the best course of action in response to Flanders' proposal would be to dissolve the Wayne Automobile Company. They then would form a new company whose name would be publicly associated with three of the most well-known personalities in the industry at that time: Everitt, Metzger, and Flanders. In turn, the new concern would take over the Wayne Automobile Company properties and use them as the basis for building a smaller, lower-priced touring car.[3-19]

Chapter Four

EMF Bursts onto the Automotive Scene

The time: Tuesday evening, June 2, 1908.

The place: The Café des Beaux Arts in the heart of New York City.

The attendees: The most influential automotive writers in New York and their wives.

The purpose: Dinner, of course, and the formal announcement of the birth of a powerful new automobile firm—the Everitt-Metzger-Flanders (EMF) Company.

United in this venture were three of the most well-known men in the Detroit automobile industry: Barney Everitt, head of the Wayne Automobile Company; William Metzger, past sales manager of Cadillac and one of the founders of the Northern Motor Car Company; and Walter Flanders, who had recently resigned as head of production for the Ford Motor Company to take the same post at the Wayne Automobile Company. For years to come, the three men would be known as "The Big Three."[4-1]

Add LeRoy Pelletier

The fact that the wives of the automotive writers had been invited to accompany their husbands to this dinner was most unusual. It was the inspiration of E. LeRoy Pelletier, the advertising manager of the company, who noted that writers' wives normally are not invited to attend such business functions but that he had "discovered that some men in New York really love their wives," hence the departure from the norm.[4-2]

LeRoy Pelletier was very good at coming up with such gimmicks. He had quite a remarkable background. A trained engineer with a knack for words, he had been a reporter who wrote about the gold strike in the Klondike. He then had helped found the Duquesne Motor Car Company of Buffalo, New York in 1904, but had resigned as its president in 1905.[4-3] Pelletier then gravitated toward Detroit and found employment with the Ford Motor Company as Henry Ford's private secretary and the Ford Motor Company "publicity engineer," otherwise known as advertising manager. Pelletier coined the popular phrase, "Watch the Fords Go By." At some point, he left Ford Motor Company to return to Buffalo to head the advertising department of the Maxwell-Briscoe Motor Company, then switched gears again to join the fledgling EMF.

Aside from the announcement of the formation of EMF, the dinner revealed some unusual asides, especially with regard to the identities of several of the company in attendance. The master of ceremonies was none other than Benjamin Briscoe, president of the Maxwell-Briscoe Motor Company. Briscoe apparently bore no ill will for the loss of his advertising head.

At the Briscoe table sat E.P. Chalfant, assistant manager of the Association of Licensed Automobile Manufacturers (ALAM), and Alfred Reeves, manager of the American Motor Car Manufacturers' Association (AMCMA). The ALAM and AMCMA had been bumping heads for several years. The ALAM had obtained ownership of the Selden patent and had been using every legal strategy at its disposal since 1903 to force all auto manufacturers to pay a royalty to it for every car built. A counter-organization, the AMCMA, whose most vociferous and stubborn member was Henry Ford, had come into being in 1906 in opposition to the ALAM. To see Chalfant and Reeves breaking bread together and joining in friendly conversation in itself was a newsworthy event. Reeves subsequently jumped ship in 1910 to take over the ALAM. This brought about the downfall of the AMCMA. Ford and other non-ALAM members, however, won the Selden court suit in the end, the judgment of which was awarded in 1911. The following year, the ALAM— no longer a power in the industry—passed into history. However, that passing would be a few years hence.

EMF Meets the Press

The announcement that EMF would be capitalized at $1 million probably came as no surprise considering the people involved. However, no product was shown, although the claim was made that a pilot model of a four-cylinder, 30 horsepower vehicle called the "Everitt" would be on the road by July 1. It would have a wheelbase of 103 inches and be in production in various body styles by October 1, 1908. The stated intent was to produce 12,000 Everitts by September 1909. This goal must have caused the raising of more than a few eyebrows because Ford had been the only company at that time to have produced more than 10,000 automobiles in a single season. However, because the man who had made that Ford production figure possible—namely, Walter Flanders—was behind the EMF operation, then why should not the new company also be able to reach such a prodigious level of output?

The formation of EMF probably caught journalists and the auto industry off guard. Rumors had abounded for several weeks that the Wayne Automobile Company and the Northern Motor Car Company were about to merge. However, no merger was announced. Instead, the news that came out was the founding of a new company, the Everitt-Metzger-Flanders Company, whose officers were Barney Everitt, president; Bill Metzger, secretary; W.T. Barbour, vice president; Charles Palms, treasurer; and Messrs. Book, Bennett, and Flanders, directors. Flanders was slated to be the general manager.

One could not help but wonder how puzzled the journalists may have been by the people involved in the coming out party of EMF. First, the principals involved still were running other car companies. Everitt was president of the Wayne Automobile Company, Palms was its treasurer, and Book was one of its major stockholders, with Flanders being its head of production. Likewise, there was William Barbour, not only the reigning head of the Northern Motor Car Company but also, with Bill Metzger and Albert Bennett, one of its largest stockholders. The only reference to the Wayne and Northern companies was that the new EMF "Everitt" would be built in the Wayne Automobile Company factory and in two plants belonging to another Michigan concern "in which Mr. Metzger is a major stockholder."[4-4] (Northern had factories both in Detroit and Port Huron, Michigan.)

Obviously, a host of questions had to be resolved. It did not help matters when on the day of the dinner, June 2, in response to a question put to the Northern Motor Car Company by the magazine, *Horseless Age,* about the possibility of a merger, the reply was, "We have not consolidated with the Wayne [Automobile] Company. Please make no such announcement."[4-5]

As events subsequently played out, some of the statements made by EMF about its plans and products during the dinner were borne out, and others were not. This was no reflection on what originally was told to the press; rather, it was merely the fact that the new company had to make adjustments over the next few months as the harsh reality of creating a vehicle, finding a place to build it, and formulating a business plan had to be faced.

First and foremost, true to their words, the EMF management had its pilot model in hand by July 1908.

EMF's First Born: The EMF 30

The model upon which EMF would pin all its hopes was the intermediate-sized, intermediate-priced EMF 30. Architect of the EMF 30 was the enigmatic William Kelly, which probably surprised no one inasmuch as William Kelly also was chief engineer for the Wayne Automobile Company where the EMF 30 pilot model was engineered. Despite the announcement of the formation of EMF in June, there is question whether it as yet legally existed because the original date of its incorporation is listed as August 4, 1908, as recorded in its first annual report to the state of Michigan. This meant that any early work on an EMF product probably was conducted under the auspices of the Wayne Automobile Company.

William Kelly: EMF Mystery Man

William Kelly, who had engineered all Wayne Automobile Company products since the company was founded in 1903, had extremely close ties with Barney Everitt and Bill Metzger. These ties have never been adequately explained. Company literature—whether of the Wayne, EMF, or later the Metzger Car Companies—tended to place Kelly on the same pedestal as a Henry Ford in terms of his engineering background.

In an article that appeared in the *Cycle and Automobile Trade Journal* in 1909, Kelly was described as[4-6]:

> ...one of the pioneer gas-engine experimenters of Detroit. He worked with Henry Ford on his early models and designed the successive models of the Wayne and the EMF of 1909. He constructed what was, in all probability, the first planetary transmission ever assembled, and now holds many patents.

Kelly did claim that he had been working on the development of a gasoline-powered vehicle in 1895, roughly around the same time that Henry Ford began to engineer his Quadricycle. However, Kelly had abandoned the effort because it was not meeting his expectations. Kelly also claimed that he had returned to his experiments in 1901.[4-7] In all probability, the latter effort is what culminated in the design of the first Wayne automobile. The fact that the first Wayne and the first Ford both had planetary transmissions might suggest a connection between Kelly and Ford. However, this would be begging the question inasmuch as the curved dash Olds also featured a planetary transmission, and it predated the Wayne and Ford automobiles by two years.

An early EMF advertisement extolled the virtues of William Kelly in grandiose terms:

> Chief Engineer Wm. E. Kelly has been in the business from its infancy—he antedates some of the "pioneers" who make greater pretensions [Ford?]. Like others he has designed some cars that did not come up to his expectations—in later years by riper knowledge he has produced some of the very best. If he had a fault, it was in aiming too high. And he has always been handicapped by having to work under the "assembling" system. In the E-M-F. "30," Kelly and his corps of designers have realized a long-cherished ambition.

A Bird's Eye View of the EMF 30

On the heels of such words of praise, we would expect the EMF 30 to be a blockbuster of a vehicle—one that would truly extend the envelope of vehicle technology for that time. Curiously, a bit further in that same advertisement, we find the following words (the italics are mine):

> Nothing added—no frills or furbelow. Nothing omitted that experience has proven or convention taught you to consider a necessary part of a first-class motor car. *Not one original feature—not a single novelty—no startling innovations. Not one experiment—not one hair-brained theory or half-baked mechanical idea—not an untried or unproven invention—or metallurgical hallucination—will you discover in the E-M-F "30."*

Obviously, Kelly and his corps were taking no chances.

Despite these disclaimers, the EMF 30 offered as much as, if not more than, other cars of its class, such as Buick, Cadillac, Oldsmobile, Maxwell, or Regal. Reports on the technical features of the EMF 30 already were being highlighted by the automotive press toward the end of July 1908. *Motor Age* called it "one of the sensations of the summer," and that "not many were prepared for so many features and novelties."[4-8] The general consensus seemed to be that the company was introducing a superior product at a price much lower than would be expected.

The engine developed by Kelly was a four-cylinder unit with a four-inch bore and four and one-half-inch stroke, capable of 30 horsepower. The pistons were five inches long with four eccentric compression rings each. Connecting rods of drop forged steel had an I-beam cross section. Mechanically operated valves extended along the left side of the block. The valves were rather large in size, being two and one-half inches in diameter and made of drop forged, special steel. The crankcase also was made of drop forged, special steel and mounted on three main bearings of babbitt. Lubrication was via the splash system with the oil contained in a special oil reservoir that was cast integral with the crankcase. Its supply was good for 300 to 500 miles before needing replenishing. The carburetor was conventional—a float-feed design with a single jet—but one built by EMF for its own products.

One of the more unusual features of the EMF 30, at least for a vehicle of its price, was its double ignition system made up of a magneto backed up by an

EMF Bursts onto the Automotive Scene

Front and rear views of a 1910 EMF 30.

emergency system consisting of a battery coil and commutator. The magneto was installed within an oil-proof, dust-free case; the battery coil and commutator ignition had no exposed leads so that the system was quite waterproof. Engine cooling was based on the thermo-syphon principle, assisted by a belt-driven fan mounted on the motor instead of the radiator.

The EMF 30 transmission was a three-speed plus reverse sliding gear type with standard gear ratio of 3.25:1. Gear sets of 3:1 and 4:1 were optional.

(The company claimed that with the higher gear ratio, the EMF 30 could achieve a speed of 50 miles per hour.) Instead of the usual square gear shaft, the EMF unit was round and had four keyways milled out of its surface. The center of each gear also was ground to improve alignment and thereby lead to more silent engagement. The propeller shaft was equipped with two universal joints. Bevel gears made up the differential—a practice more common on expensive cars—again to achieve more silent operation. The sliding gears, propeller shaft, and the bevel driving gears were of special alloy steel. Integral with the transmission case was a semi-floating live rear axle constructed in such a way as to create an axle lighter and stronger than that found on other automobiles of similar power and weight. All gears performed within an oil bath contained in a tight housing designed to prevent leaking. The clutch was an expanding ring type with leather facing.

Other chassis details include an I-beam front axle and irreversible worm and sector steering gear. Braking was accomplished through a pair of external contracting bands lined with camel's hair acting on pressed steel drums at the rear wheels. Emergency braking came from a metal band expanding against the inner face of the same drums. Pressed steel discs surrounded both drums to keep them dust-proof.

The EMF 30 frame was of U-shaped, pressed steel construction and rode on semi-elliptic springs at the front and full elliptic springs at the rear. Wheels were of the 12-spoke artillery type and carried 32 by 3.5-inch Morgan & Wright tires on quick detachable rims. The wheelbase was 106 inches, with a tread of 56.5 inches. Weight (of the touring car) was 1,800 pounds, and the gasoline tank had a 15-gallon capacity. The body came in four iterations: touring, runabout, rumble seat roadster, and demi-tonneau (i.e., runabout with a detachable rear seat). The standard color was red.[4-9]

It seemed to be generally accepted that the EMF 30 offered a modern, high-quality car at a price ($1,250) much lower than would normally have been expected for a vehicle with its features. This was possible because the EMF cars were not assembled cars (so the rationale goes) but used parts made by the parent organization. Underlying this rationale was the undeniable fact that production was in the hands of Walter Flanders, whose genius on the plant floor could not help but ensure that the EMF 30 would be a product of the highest quality at the lowest cost.

A catalog advertisement for the 1910 EMF 30 touring car, which sold for $1,250 at that time.

It's Not the Product, But the Process, That Counts

There is no doubt that the Flanders name helped bring enormous interest to the EMF 30. Furthermore, whatever the EMF 30 may have lacked in product innovation was more than compensated in the process of manufacture.

Later, in an advertisement regarding its upcoming 1911 EMF models, Flanders boasted that by virtue of his ability to achieve large-quantity production, he had fulfilled his promise to provide the public with an automobile of higher quality and lower price than any of his competitors. Moreover, he had achieved this elevated status by virtue of the ability of EMF to manufacture every component of the car within its own factories. It could forge, stamp, and heat-treat its own steel; cast its own cylinders; and even make its own bodies. In this sense, Flanders predated Henry Ford whose dream, realized later in the Rouge River complex, was to be entirely self-sufficient in the manufacture of the Model T.

When we read the pronouncements of Flanders in advertisements or in newspaper and magazine reports of that day, we have the eerie feeling that it is Henry Ford who is doing the talking, not Walter Flanders. Flanders continually stressed his desire to build a car that would meet everyone's needs and would be so perfect and long lasting in all respects that only minor changes would be necessary as time passed.

Similar to the Ford philosophy of later years (or possibly even of that time), Flanders reasoned that by manufacturing all his own parts, he could omit the middleman, or the parts supplier. This would allow him to buy materials in such large quantities that he could bargain for cheaper prices than could his competitors and then pass the savings on to the customer. Indeed, most cars of that era were "assembled" cars, and companies depended on suppliers to manufacture their significant components.

Flanders Revolutionizes the Production Process

The genius of Walter Flanders was that he had the vision to recognize that building a complex piece of machinery such as an automobile could be accomplished in much the same manner as the building of other, more mature products of mass production such as bicycles, guns, stoves, and sewing machines. The task was much more difficult, but it could be done.

One of Flanders' greatest contributions was that of eliminating the human element in performing factory tasks where the accuracy and uniformity of a given assembly could make or break the reputation of the entire vehicle in terms of quality and reliability. Flanders introduced what came to be known as the "jig job." This revolutionary practice meant that "each part on which a mechanical operation was to be performed was brought to the operation tightly clamped into a certain frame which corresponded in its dimensions to the base of the tool which was to perform the operation."[4-10] Setting up a "jig" initially added extra costs to the product, sometimes several thousand dollars; however, when those costs were amortized over a year of production that numbered in the thousands of units, Flanders forecast correctly that the costs would be well worth the investment and in time would lead to cost efficiencies.

As when he remodeled vehicle assembly operations at the Ford Motor Company, Flanders' consummate desire was to increase production beyond the norms of the day, especially by substituting machines for manpower wherever possible and having those machines perform many operations simultaneously. Flanders introduced millers that could smooth two or more faces of a casting at the same time, and spindle drills that could drill more than one hole into a casting in each operation. He brought in automatic screw machines, gear cutters, and die presses, many of which had never been seen on a conventional automobile factory floor and only on limited basis within the manufacturing facilities of parts suppliers. Such machines could do much more work in a given time than could be done by human hands, and could do it in a manner that would enable the car buyer to purchase a replacement part later in time that not only would fit but would do the necessary work.

These EMF 30 chassis are ready to receive their bodies, circa 1909. (Courtesy of the Detroit Public Library, National Automotive History Collection)

Flanders' farsightedness extended beyond the factory floor. He established a chemical and physical laboratory. To manage it, he hired the best metallurgist he could find. Often, Flanders would know more about the makeup of a raw material he was purchasing than the supplier knew. The laboratory saved Flanders from spending time in attempting to judge the worth of a supplier by reputation or other means.[4-11]

So confident were the EMF backers in Flanders' abilities that when he later proposed to purchase a bathtub company on the Detroit River in which to press steel parts, there was no dissension. Apparently, it became the first plant ever in the industry installed with dies and presses dedicated solely to automobile parts production. Soon the factory was turning out fenders, hoods, running boards, body panels, and fuel tanks. At times, it also would be used to produce small parts such as hubcaps, radiator caps, and rear axle housings.[4-12]

The Preliminaries End: EMF Is Incorporated

On August 4, 1908, six influential Detroiters gathered within a red brick building on the corner of Piquette Avenue and Brush Street, the offices of the Wayne Automobile Company, "a stone's throw" up the block from the Ford Motor Company. They were: the millionaire Detroit real estate dealer, Charles L. Palms; his wealthy ex-dentist brother-in-law, James B. Book; the ever-smiling body man, Byron F. Everitt; super car salesman, William E. Metzger; machinery and production expert, Walter E. Flanders; and the mysterious and enigmatic William Kelly. Their mission was to make official what had been announced at the grandiose New York dinner of June 2; namely, the formation of the Everitt-Metzger-Flanders Company. Thus far, the building of a pilot model and the grubby details of putting together an organization were being conducted under the auspices of the Wayne Automobile Company. This was about to change.

The Wayne Automobile Company Becomes EMF

The first and foremost order of business was the creation of a memorandum of agreement between the Wayne Automobile Company and the emerging Everitt-Metzger-Flanders Company, outlining the steps by which the dissolution of the Wayne Automobile Company would occur and be replaced by

EMF. This document is extremely interesting from the standpoint of the details it gives for the creation of a new organization from an old one, and for stipulating the monetary arrangements set forth for the principals involved.

The memorandum begins by specifying that a new corporation (EMF) would be formed pursuant to the laws of the state of Michigan, and it would be capitalized at $500,000 (unlike the June 2 announcement that it would be $1 million). In turn, EMF would purchase all assets of the Wayne Automobile Company except those automobiles still unsold or in partial assembly. Rights to the "Everitt" (the early name given to the pilot model of the EMF 30 which was built under Wayne auspices) would be transferred with all other assets. With the transfer of these assets, the Wayne Automobile Company would be automatically dissolved. Charles Palms acting as trustee for the stockholders of Wayne Automobile Company would make the transfer official as soon as possible. The purchase price of Wayne would be $191,801.75, to be paid to Wayne stockholders in the form of capital stock in the new EMF company. The purchase price was the result of an inventory made by the Lemly Appraisal Company, a reputable Detroit appraiser.

The memorandum also stipulated that James Book and Charles Palms personally would assume any outstanding debts that belonged to the Wayne Automobile Company so that the assets of Wayne would go to EMF free of any monetary encumbrances. Moreover, Book and Palms would be asked to loan EMF the sum of $25,000 each for one year at five percent interest and extend a line of credit of $200,000 to EMF over a period of two years. In addition, William Metzger and Byron Everitt each were required to purchase 2,500 shares ($25,000) of EMF stock for cash immediately, and subscribe for 1,000 more shares ($10,000) to be paid for by October 10, 1908. Reading between the lines, one might assume that the Metzger/Everitt stock purchases and the Palms/Book loans would be the only ready cash available to EMF at the beginning of its business life.

Further in the memorandum, the fine print indicated that 16,250 shares ($162,500) of unauthorized capital stock would be set aside and reserved for Everitt, Metzger, Flanders, and Kelly, and would be paid for out of the dividends of the company. This stock would be issued to the four parties periodically over time as payment for their services. We might refer to them as bonuses. Considering the value of the dollar in 1908, the potential worth of the reserved stock that Messrs. Everitt,

Metzger, Flanders, and Kelly could earn was no mere pittance. Of the 16,250 shares, 6,250 shares were earmarked for Flanders, 3,750 shares each for Everitt and Metzger, and 2,500 shares for Kelly.

In reality, this plan represented a marvelous technique for retaining the services of the more valuable members of the EMF team. It provided them with an enormous incentive to make the company succeed. The allocation of the lion's share of unissued stock to Flanders reflected the fact that he would be conducting the day-to-day operations of EMF, and a good part of the future success of the new company would rest on his shoulders.

If any one of the four left EMF before June 1, 1913, the person would forfeit the right to his shares of the unreserved stock.

The memorandum also entered into the sacred domain of salaries. For example, Metzger was to be paid $6,000 per year as sales manager. Walter Flanders, as general manager, would receive a similar amount, as would Byron Everitt, although the latter's job description "to perform such general duties as may from time to time be assigned to him by its Board of Directors" seemed a bit vague. Their contracts would run for five years. Also, each of the three would receive five percent of the net profits of the company during this period, this in addition to any bonuses they received from the reserved stock holdings.

William Kelly, the mechanical engineer, would receive a salary of $5,000 per year, as would Charles Palms, acting as treasurer. If, after two years, the net profits of EMF could not sustain these salaries, they would be renegotiated.

Finally, the memorandum stated that a board of five directors consisting of John Book, Charles Palms, Byron Everitt, William Metzger, and Walter Flanders would manage the new company. Board members also would serve as officers of the company: Everitt as president, Book as vice president, Palms as treasurer, Metzger as secretary and sales manager, and Flanders as general manager.

Thereupon the EMF stockholders (who were the five men who made up the board of directors) unanimously adopted the memorandum of agreement that had been drawn up, which included the purchase of the Wayne Automobile

Company. The new officers of EMF were selected with no change from the memorandum. It now became official. The Everitt-Metzger-Flanders Company came into being on August 4, 1908.[4-13]

Flanders Lays It on the Line: Survival Depends on High Production

As delighted as all the directors seemed to be in the formation of their new company, one officer did have some serious misgivings—Walter Flanders, the realist of the group.

Flanders had been hard at work making a cold calculation of what it would take to make EMF a successful company. He estimated that to equip a factory to build the EMF 30 and place it on the market would require all the assets, building, and lands of the Wayne Automobile Company, roughly $95,000 in cash, and a line of credit from the bank for approximately $500,000. Furthermore, his calculations had taken into account that the Wayne Automobile Company property would be free and clear of any debt. He had not anticipated that EMF also would be required to assume the liabilities of the old Wayne Automobile Company or the marketing of the remaining Wayne cars that had not yet been sold. (Palms and Book apparently had insisted on this change in plans.) These further obligations—namely, to wind up the old Wayne Automobile Company affairs, eliminate its debts and charge off its worthless accounts, and sell the remaining Wayne vehicles—severely intruded upon the time that Flanders was able to devote toward placing the EMF 30 into production. Nevertheless, he persevered and, to his credit, actually realized a profit of approximately $30,000 for EMF in doing so.[4-14]

Flanders' projected build of 12,500 EMF automobiles by the end of 1909 was sufficient to set the entire industry back on its heels. It was an outlandish projection for those times, especially for a car that sold within the intermediate price range. Only one manufacturer had ever produced more than 10,000 automobiles thus far in a single year, and that was the Ford Motor Company in 1907. (That number also might have been no more than 7,000, depending on one's sources.) The man who made that Ford production possible was Walter Flanders, the same Flanders who now governed the production of EMF. If he had succeeded once before, why should he not be able to reach such rarified air with another company?

The E-M-F Company

The original Wayne/EMF factory (far left), as it appears today. The narrow white front of the Ford Piquette plant is shown at the end of the next block.

Another view of the EMF factory (shown in the center and left of this photograph), as it appears today.

Unfortunately, there was the problem of distribution. Even if Flanders did succeed in building 12,500 EMF automobiles through the following year, how was he to market them? The Wayne sales force that had been absorbed by EMF had never handled more than 600 cars in a single season. To ask it to market 20 times that number in a single year was completely out of the question.

Enter Frederick Fish and Studebaker

Fortunately, an answer to the distribution and sales problem dropped from the sky as a rescuing angel: the Studebaker Brothers Manufacturing Company. Frederick Fish of Studebaker had heard of the formation of EMF. He also was quite aware of the talents of Everitt, Metzger, and Flanders, and he was itching to broaden the toehold of his company within the automobile industry. Fish had made contact with Flanders to discuss a proposal that could be of mutual benefit to both companies; namely, that Studebaker become the sales arm of EMF. Flanders definitely was interested because it would relieve EMF of the expense of setting up a national sales force.

On July 8, Messrs. Frederick Fish, Clement Studebaker, Jr., Nelson Riley, George Studebaker, and Hayden Eames representing the Studebaker board of directors visited the Wayne/EMF works to explore firsthand the possibilities that lay open to them. More specifically, they wanted to examine the EMF 30 and judge its merits before they decided to allow it to be sold in Studebaker wagon outlets, which already stocked the expensive Studebaker-Garford and electric models. Would the addition of an unknown brand at a moderate price have a negative effect on the sales of their more luxurious models? (The question really was moot because sales of the latter were so few that it is difficult to imagine that anything could alter their availability.)

The Studebaker group returned to South Bend and reported their findings at a special meeting of the board of directors on July 10. Speaking for the group, Frederick Fish said that they had found the EMF 30 to be of merit. He posed two questions to the board: (1) Should Studebaker take on the sale of an intermediate-priced car such as the EMF 30? and (2) Should they contract for the first lot of cars built by EMF before January 1, 1909, pay for them as they are sold, and then pay 50 percent of the cost of all models delivered thereafter? The board gave its affirmative response to both questions, especially making the point "that the cars need only be ordered as sold or as they are wanted for use at the branch houses of Studebaker Brothers Manufacturing Company."[4-15] Most curious was the fact that Studebaker already had a similar arrangement ongoing with Arthur Garford, as a result of an earlier attempt to enter the auto industry.

Studebaker's Earlier Attempts to Enter the Automobile Age

Studebaker, the largest wagon maker in the country, had been flirting with the automobile business since 1897 when it contracted to build 100 taxicab bodies for the New York Electric Vehicle Company. Late in 1902, the company began to build a smattering of electric vehicles, their chassis being supplied by Arthur Garford. The two companies then agreed to build the expensive Studebaker-Garford gasoline-powered car.

The Affairs of Studebaker-Garford

Arthur Garford was an unusual study. He had made a fortune in bicycle seats and then, similar to Metzger, decided that the automobile was on the threshold of great things. He was determined to be a part of them. His first attempt was to form a joint venture with the Pope Manufacturing Company, a business called the Federal Manufacturing Company. George Pope was the leading bicycle manufacturer of that day. He too had turned toward the automobile as the bicycle craze began to wane. In 1904, Garford began to suspect that Pope wanted to push him out of their concern, so he terminated their relationship. He then bought out Pope's interest in Federal and reincorporated the company.

One month later, Garford was approached by George Studebaker, whose company had a new gasoline-powered vehicle on the drawing board. Studebaker asked if Garford would be willing to supply the chassis for its automobile. Garford agreed, albeit reluctantly, but because he already was building the chassis that Studebaker used in its electric cars, he decided that he had little to lose.[4-16]

By 1906, Garford was building approximately 500 chassis per year in his Cleveland plant, most of them for companies other than Studebaker because the Studebaker-Garford car sales were so few. Garford now began to adopt illusions of grandeur. He desired to build his own car—a high-powered, intermediate-sized car that would carry his name alone.

In the spring of 1906, Garford recapitalized his company, with the expectations of raising more money so that he could build a new plant in Elyria, Ohio. In this plant, he expected to build up to 1,200 Garford cars a year. He placed 6,500 new shares of stock on the market, 39 percent of which were immediately purchased by Studebaker. So large a purchase gave Studebaker the right to place three of its members on the seven-man board of Garford. Studebaker also contracted to be the sales agent for half the projected build of vehicles in the new Elyria plant.[4-17]

Studebaker's next move stunned Garford, who was told in no uncertain terms by Frederick Fish that Studebaker had first priority on all chassis that Garford built. Only when the Studebaker orders were filled could Garford think of building the chassis for any other company or for the Garford car itself.

Nevertheless, totally occupied in the build of the Elyria factory, Garford continued to ignore the red flags that the Studebaker interests had raised. No doubt he was entranced by the prospect of building his own car in 1907.

However, by May 1907, the Elyria plant still had not been completed. There were construction and machinery layout problems galore. Total disruption came when the organization was hit by a myriad of labor problems that had begun in the Garford-Cleveland plant and subsequently had spilled over into the Garford-Elyria plant.

Thus far, Studebaker had allowed Garford to conduct his own battles. However, in May 1907, the Studebaker directors accepted Frederick Fish's recommendation to reorganize and forge a merger with Garford.

Now it was Studebaker's turn to be stunned. Despite his labor and financial problems, Garford refused Studebaker's overture for merger out-of-hand.

Studebaker now girded its loins for all-out war. Initially, the company took a softer, new approach calculated as a first step that would bind the two companies closer together. Fish dangled the carrot that Garford could use the

Studebaker name along with his own on the automobiles built at the Elyria plant. He also suggested that a new sales agency be formed to represent the two companies. The agency would sell cars under both the Garford and the Studebaker names. Two months after refusing the initial offer of merger, Garford capitulated to this latest of what would be several calculated moves geared to take over his company.[4-18] He had exhausted his own resources.

Those moves were not long in coming. On February 13, 1908, Frederick Fish announced that Studebaker had become the majority stockholder in Garford. He informed Garford that from this point onward, sales of Studebaker-Garford products would take place only in Studebaker dealerships, thus casting adrift all existing Garford dealers. (Later in the year, Fish would go so far as to inform Garford that the sign over the Elyria plant henceforth should bear the title, "Studebaker-Garford.")

Five months later, despite his allegiance with Garford whose company now was apparently under Studebaker domination, Frederick Fish made his bold approach to Walter Flanders at EMF.

Studebaker Signs a Sales Contract with EMF

The outcome of the Fish proposal to Flanders came in the form of an interesting, if puzzling, announcement from both the Studebaker and EMF headquarters in mid-August. It said that Studebaker had agreed to take on half the projected output of EMF for the coming year for sale at its wagon outlets, or approximately 6,000 automobiles. Studebaker, with its huge network of five branch houses and 4,000 to 5,000 retail dealers, would be in charge of sales in the West and South, and would have exclusive control over any trade outside the United States. Bill Metzger and the EMF sales force would handle the Midwest and East. Metzger already had begun his campaign to sign up dealers, and he reported capturing applicants in some of the larger cities whose names would come as a surprise when revealed later.

A call from *The Automobile* magazine to Colonel George Studebaker elicited the following response[4-19]:

We are highly pleased at its consummation and frankly believe it will prove to be the most important move that has been made in the automobile business. We considered it more advantageous to us to form an alliance with a group of men such as that comprising the Everitt-Metzger-Flanders Company, possessing as they do factory facilities, experience, and manufacturing ability of a rare order, as well as an intimate knowledge of the problems peculiar to the motor car, than to establish a separate factory of our own.

No one, it seems, felt it necessary to ask Studebaker why that company should feel the need to link up with EMF since it already had a similar arrangement with Garford.

No doubt, the emergence of a new company such as EMF headed by such powerful figures as Everitt, Metzger, and Flanders presented a much more promising vehicle for entering the motorcar business than did Garford, who was a very small player in that industry. The contract to sell 6,000 EMF vehicles in 1909 indicated how promising that alliance could be to them, and that was where the attraction lie. It probably represented more automobiles than Studebaker-Garford had produced and sold during the entire four years of the latter relationship. (In fact, only 500 Studebaker-Garford cars were purchased during 1908, but even that small number netted Studebaker more than $1 million.) Moreover, even if the new Garford Elyria plant resolved all of its problems and went into full production, it was not capable of building more than 1,000 automobiles each year, far short of the EMF projection. Because of its more limited production capacity, Garford was necessarily wedded to the manufacture of high-priced models selling in the $3,000 and $4,000 range. The EMF 30 price of approximately $1,250 would appear to be more to its liking in terms of attracting a higher sales volume.

Why EMF Needed Studebaker

On the part of EMF, the attractiveness of the numerous domestic Studebaker sales outlets was undeniable. In a letter sent to Metzger, Everitt, and Palms dated July 29, 1908, Flanders outlined the rationale for accepting the Studebaker proposal, and attached a memorandum stating the latter's terms with the proviso that a decision must be reached by the next day. He argued that a tie-in with a company of the size and prestige of Studebaker would have undeniable rub-off effects on EMF. He added that by taking over the sales responsibility for the Southwestern and Western regions and for sales in foreign countries, as well as assuming all charges for introducing the cars into these areas, Studebaker would save EMF a considerable amount of money over the next six months. This savings was important because the company clearly needed this money to build the large number of EMF 30's that Flanders had projected. He also thought the foreign sales aspect was valuable because EMF was too young to attract top-grade foreign sales talent and did not have the funds to make a meaningful initial impact in that market.[4-20]

Flanders' understanding of the proposed sales agreement was that Studebaker would not object to EMF building a less expensive car in the future. Because Studebaker stated that it would market EMF's in the West and Southwest, he assumed this meant that EMF would control the Eastern market. He had no problem either with the Studebaker suggestion that the nameplate on EMF cars read, "E.M.F. Co. 30, built for Studebaker Bros."[4-21]

Whether EMF was aware of or concerned about the troubled Studebaker-Garford relationship is not known. On the surface, it would seem that EMF did not have such a concern at the time because later developments embroiled both Studebaker and EMF in a rather messy court suit that had all the same trappings as Studebaker's problems with Garford.

Flanders had other reasons for bringing Studebaker into the fold. They had to do with the fact that he planned to have initial production underway in October. Unfortunately, this meant that EMF vehicles would be making their first appearance on the market during the fall and winter months when automobile sales traditionally were at their weakest level because of bad weather. This would bode no good for the young company for several reasons. First and foremost was Flanders' concern that EMF at this point was operating on a financial shoestring. The company needed a healthy income from sales to plow back into manufacturing if EMF were to reach a build rate of 400 to 500 cars per month by April 1909. In Flanders' mind, this could be achieved only if EMF could build and sell at least 1,000 cars in total between October and April, which would bring the company an income of approximately $860,000. His anxiety about having Studebaker on board stemmed from his belief that the latter, with its extensive sales force in the West and Southwest, free of the bad weather of winter, could sell up to 630 to 640 of the 1,000, thus ensuring the success of EMF. Without the help of Studebaker, EMF probably could not sell more than 400 cars on its own during the same period, reducing the necessary income by more than half and making it increasingly difficult for Flanders to increase production substantially during the coming year. Furthermore, the fewer cars that EMF could manufacture through the fall and winter, the longer it would take to train workers to build them in larger numbers and free of defects later.

Whether the entire board of directors agreed with Flanders' assessment of the situation or not, it nevertheless decided to forward a contract to Studebaker regarding terms for the sale of EMF automobiles, to which Studebaker agreed by a letter dated August 5, 1908.

EMF Formalizes the Purchase of the Wayne Automobile Company

As difficult as it may be to understand, the EMF company, although having completely taken over the Wayne Automobile Company, still did not formally own it. Finally, on August 28, 1908, Walter Flanders wrote a letter to the EMF board of directors. In it, he recommended the purchase of

> ...the entire tangible and intangible assets and good will of the Wayne Automobile Company, for the sum of Two hundred fourteen thousand dollars ($214,000), payment to be made on the basis of Two hundred eleven thousand dollars ($211,000) in stock of the Everitt-Metzger-Flanders Company at par and Three thousand dollars ($3,000) in cash.

This amount was roughly $22,000 more than the figure accepted by the board at its last meeting, no doubt because EMF had given in to the request of Palms and Book that the new company underwrite the old Wayne Automobile Company debt. The letter also stated that the purchase was to be made with the understanding that EMF would not be responsible for any liability involved in three lawsuits still pending against Wayne Automobile Company (amounting to approximately $16,000) nor any future lawsuits.[4-22]

Thus, the EMF board met on September 15, 1908, and voted to purchase all assets of the Wayne Automobile Company in accordance with Flanders' letter. The board then went one step further and allocated the sum of $26,000 to add a new building to the Piquette property.[4-23] The EMF Company now was more than just a name on paper. It owned a factory and already was thinking of expanding.

EMF Buys Control of the Northern Motor Car Company

Already by September 17, 1908, the EMF stockholders (essentially the board members plus William Kelly) had drafted a letter to the directors of the Northern Motor Car Company, stipulating the terms for the latter's purchase. It proposed that EMF buy the entire assets of the Northern Motor Car Company for a sum of $200,000, the sum determined as a result of an appraisal of the property conducted by the Lemley Appraisal Company on September 14, 1908. The $200,000 would be paid to Northern Motor Car Company stockholders in the form of capital stock in EMF. A second condition was that the following stockholders of Northern were to loan EMF $25,000 each at five percent: William Barbour, Mrs. William Barbour, George B. Gunderson, and William Metzger. Half would be due on October 15 and the other half on November 15. EMF also agreed to assume all debts of the Northern Motor Car Company, which were estimated to be $185,000, except for bills outstanding.[4-24] (Of the $185,000, $150,000 was earmarked for William Barbour and Gunderson, apparently in repayment for earlier loans they had made to the Northern Motor Car Company.)

The purchase of the Northern Motor Car Company was not popular with Flanders, who felt other members of the board had forced it on EMF, particularly Metzger.

Flanders later reminded the EMF directors that[4-25]

> During this period [while cleaning up the Wayne Automobile Company's leftover affairs], and at the earnest solicitation of Mr. Wm. E. Metzger, the absorption of the Northern Automobile Company came up for discussion and negotiations for the purchase of the Northern Company's plants, together with its

> assets, extended over a period of some three months, and it entailed upon the General Manager of this Company much detailed work and responsibility which tended to hold back and impede in many ways the production of the E.M.F. 30 Model.

He added further,

> I wrote a letter to each of the directors of this Company recommending that all such negotiations with the Northern Automobile Company be dropped without further delay, and I also advised the Northern Automobile Company's stockholders to that effect.

Nevertheless, it was the will of Metzger, not that of Flanders, that prevailed with the EMF board. However, Flanders did have a heavy hand in compiling the numbers that went into the Northern purchase. For example, when the Northern Motor Car Company provided EMF with an inventory of its assets, which supposedly amounted to $546,000, Flanders reviewed it and immediately pared the list down to $411,000. He then wrote a letter to Northern dated September 17, 1908, in which he stated that EMF would buy the entire assets of the Northern Motor Car Company for $200,000 in stock in EMF, assume the debts of the latter company, and sell off its 250 remaining automobiles. In what may have seemed to be an afterthought, Flanders added that these conditions also were dependent on the Northern stockholders loaning EMF the sum of $100,000![4-26]

Flanders made certain that the EMF board members harbored no illusions about the status of the Northern Motor Car Company, which he called a "rather poor business proposition."

The intricacies of the arrangements by which the Wayne Automobile Company became EMF, then purchased the Northern Motor Car Company, give vivid example of the financial adventuring that took place among early automobile companies. One has the feeling that the management of car firms in Detroit during this period met periodically, perhaps for lunch or dinner at the Pontchartrain Hotel, to exchange notes and be supportive of each other.

The Northern Buyout Is Made Official

In October 1908, when the EMF board of directors and stockholders met again, the Northern Motor Car Company purchase was, for all intents and purposes, all but decided. First, one significant matter had to be resolved: to increase the capital stock of EMF so there would be enough shares to trade for the Northern Motor Car Company. The EMF board therefore authorized the directors to prepare the papers and certificates necessary to increase the capital stock from an aggregate of $500,000 to $1 million (100,000 shares at $10 par). Of this amount, $193,240 (19,324 shares) would be issued to Northern Motor Car Company stockholders in full payment for the purchase of their company. William Barbour, Mrs. Ellie T. Barbour, George B. Gunderson, and William Metzger, whose holdings of 5,000 shares each or almost half the outstanding Northern shares in total, were the main recipients. With 40,250 shares outstanding, the sale represented a single share value of $4.80.

By acquiring the Northern Motor Car Company, EMF added two factories in Detroit and one in Port Huron to its spreading production facilities. Flanders explained that the new plants would allow him to put into practice a plan that had been on his mind for some time, which was to have a "separate factory for the production of replacement parts for all previous models of Wayne and Northern cars, as well as for all future E-M-F models."[4-27]

Flanders Moves to Protect Ex-Wayne and Northern Car Buyers

Flanders evidently later changed his mind about his use of the Port Huron factories. The reality was that he approached a close friend, Alfred Owen Dunk, and persuaded him to assume the responsibility for supplying replacement parts to Wayne and Northern customers. Dunk, who then was president of the Puritan Machine Company, honored his friend's request and organized a second firm, the Auto Parts Manufacturing Company. He then purchased from EMF all existing parts, drawings, jigs, patterns, and special tools that would enable him to make, if necessary, any part that a previous Wayne or Northern buyer might require. He also brought in L.A. Austin to help manage the business. It was a wise choice, inasmuch as Austin, formerly an assistant manager at Northern, had been relegated to the duty of responding to parts requests for Wayne and Northern automobiles when he moved to EMF.

Dunk thus was handed a ready-made business. Being a rather enterprising young fellow, he went on to make a tidy fortune for himself as time passed by buying other automobile companies when they went bankrupt, so he could be the supplier of their replacement parts as well. In fact, he became so successful that he also converted his other company, the Puritan Machine Company, into the replacement parts business. Eventually, he took over and ran the Detroit Electric Car Company, the last of the 36 electric car manufacturers to have existed in the United States.[4-28]

Alfred Owen Dunk, an enterprising man who earned his fortune as owner of the remains of 756 automobile companies. (Courtesy of the Detroit Public Library, National Automotive History Collection)

In the April 20, 1929 issue of *Automobile Topics*, a small but rather astonishing item appeared. In it was reported that Alfred Owen Dunk had "presented to the National Automobile Chamber of Commerce all the original tracings, drawings, blueprints, and United States and foreign patents which were the property of some 756 automobile plants which have gone out of existence." Dunk had accumulated the paper property of that many companies in two decades. It all began with the concern of Walter Flanders that buyers of Wayne and Northern cars would continue to receive service after the two companies that built those cars had been absorbed by EMF.

Struggling to Get Off the Ground

With one thing leading to another, Flanders was having a difficult time achieving the volume of production that he had forecast. By the end of the year, only about 172 EMF 30's had been built. Worse yet, each one had to be recalled because the thermo-syphon system it was using instead of a water pump was not up to the task.

Ignoring problems on the production end, Metzger took great pains in having the EMF 30 unveiled at the January 1909 New York Auto Show. It received excellent reviews in leading industry journals such as *Automobile* and *Motor Age*. *Automobile*, for one, made it a point of dwelling on the large diameter of the valves found in the EMF 30, which "are over half the diameter of the cylinders and...in motors of this make the valves are very large indeed." The write-up then praised the EMF product[1-29].

> At $1,250, this 'Thirty' is one of the surprises of the year, particularly if account is taken of the fact that it is in every way a standard tour-car, both in point of size and utility, even including a magneto in the ignition system.

At the Chicago Auto Show the following month, Metzger publicly admitted that EMF was off to a slow start, building only 20 cars per day which, if continued, would result in the company producing only half the annual number of cars that Flanders had predicted. Metzger was quoted as saying,

> We have been building new plants, which has taken a lot of our time. We have doubled the size of the old Wayne plant in

EMF Bursts onto the Automotive Scene

EMF automobiles on display in 1909. (Courtesy of the Detroit Public Library, National Automotive History Collection)

> our new building and now have built two buildings, 490 by 110 square feet and three stories high. Soon we will build an arch connecting the two buildings, which will form a square of 490 feet.

Metzger also gave another reason for the slow start[4-30]:

> ...that it is due to our determination to have our cars right, and we closely scrutinize each machine before it goes to market.

Be that as it may, EMF matters rapidly were coming to a head, culminating in March 1909 with a verbal explosion in the EMF boardroom.

Chapter Five

EMF Loses Its "E" and "M"

March of 1909 strolled into the EMF boardroom neither as a lamb nor a lion, but would prove to be a pivotal month for the organization. Barney Everitt was on the hunt, and he was using a very large caliber of gun.

Everitt, the company president, to put it mildly, was upset.

Unspoken Studebaker Contract Terms Anger Everitt

The degree to which he was upset came spilling out via a letter Everitt wrote to other EMF directors four days before their March 8 board meeting.

First, Everitt had had it with Studebaker. What particularly irritated him were the terms of the original sales contract between EMF and Studebaker. Apparently in their haste to bring the latter company on board as their western and southwestern sales agent, the EMF directors had neglected to include language that would require Studebaker to accept a specific number of cars during the terms of the agreement. Furthermore, there was nothing in the contract to force Studebaker sales agents to give EMF a deposit on the vehicles they received, which obviously eliminated a major incentive for the vehicles being sold as quickly as possible. The cars could sit on the Studebaker lots forever before EMF received payment. The fact that EMF had absolutely no control over the Studebaker sales agents with regard to their product rankled Everitt to no end.

Even more irritating to Everitt was the fact that although EMF had within its own ranks Bill Metzger, one of the top sales executives within the industry,

Metzger's hands were tied when it came to promoting sales in territories covered by Studebaker.

Everitt wrote in his letter to the board,[5-1]

> Regarding the sales end of our business, the writer has every confidence in our sales manager, Mr. Metzger, to market successfully, and that he, better than anyone else could market our entire output, no matter how large we make it, and until our sales manager has shown himself incapable of doing it, I believe the sales end of the business should be left in his hands.

Everitt continued by writing that the board should "decide once and for all that we are through with the Studebaker outfit." By his comments, Everitt did not mean that EMF should sever its sales agreement with Studebaker immediately, but should close that aspect of the contract that said Studebaker would handle the sales of all EMF cars after August 1, 1909.

If that did not suit the board, Everitt offered an alternate proposal:

> ...there is but one way that they [Studebaker] can have the output of the Everitt-Metzger-Flanders Company for 1910, and that way is that they buy outright for cash, the entire organization.

This was rather heady stuff, calculated of course to spur the board to take action. It does not appear that Everitt honestly believed that the latter alternative would be accepted. It was used more as a tool of argument to sway the other board members to end their relationship with Studebaker.

Everitt took a calculated risk, and there is good reason to believe that it backfired on him, not so much because the board did not follow his reasoning but because of other matters contained in the March 4 letter that temporarily may have affected his relationship with the powerful Walter Flanders.

Everitt Is Upset with Flanders Also

Everitt suggested that Walter Flanders was spending too much time on matters involving the EMF relationship with Studebaker, to the detriment of his parent company. Moreover, it appeared that some defects had been discovered in EMF automobiles—defects that had to be resolved immediately to maintain the high reputation of the young company. Everitt went on to say,

> This is a very serious stage in the career of this new company, and the defects which have developed in the car, small though they are in themselves, may produce very serious results unless followed up and taken care of. They call for all the time and thought of our General Manager [Flanders], and the Company needs his loyalty today more than it ever will at any other time.

Coupled with this mild rebuke was a proposal that all complaints about EMF automobiles should be handled initially by the EMF sales department (i.e., Metzger), not the engineering department. Everitt felt quite strongly that it was the responsibility of sales to maintain close contact with the customer to ascertain if any problems might exist or if other needs could be satisfied, sometimes with little effort. Sales then would inform engineering as to what the response should be. Although this was an admirable point of view for such a young company, it could not be placed into effect by EMF because of the sales contract with Studebaker. Studebaker agents, not those of EMF, were at the point of sale for two-thirds of the EMF output. The Studebaker agents owed their loyalty to their parent company, not EMF. What did they care if they received complaints about the cars they were selling when the primary products for which they were held accountable were wagons?

Everitt had a good point. Could the Studebaker sales agents be trusted to push the sales of EMF automobiles as enthusiastically as they pushed the sales of Studebaker wagons? He doubted it.

> We know but very little of the real conditions in the territory handled by the Studebaker Company, but I believe that the Everitt-Metzger-Flanders Company should have a personal investigation of every Studebaker dealer in the West, and

> while this would entail considerable expense, it would be money well spent; that the Everitt-Metzger-Flanders Company should start immediately to lay lines for the marketing of its product in the territory handled by the Studebaker Company at once, and not leave ourselves at the mercy of any outside company.

The language of the letter further indicates that Everitt might have been miffed that Flanders arbitrarily made decisions involving company (not necessarily production) matters without consulting with the other officers, especially the president [5-2]:

> While the writer has absolute confidence in our General Manager's ability to manage, he also thinks, and will insist, if the writer remains connected with the organization, that all matters of importance be brought before at least two members of the board, and would also insist that he, as president, be one of those members.

Everitt's argument had merit because the company by-laws limited the authority of the general manager insofar as he could not transact business that would affect the interests of the stockholders or the future life of the company. However, in being so blunt about stating his point of view, Everitt no doubt should have expected a corresponding rebuttal from Flanders. He eventually received that rebuttal in no uncertain terms, but at a later date.

Initially, it appeared that Walter Flanders took the contents of Everitt's letter gracefully because it was he who moved that the board accept them and have them included in the minutes of its March 8 meeting. The board even went so far as to act on the Everitt recommendations. By a unanimous vote, it passed a resolution instructing the president to tender an option involving the sale of all assets of EMF to Studebaker, inasmuch as the two companies could not come to any agreement over the sale of EMF cars after September 1, 1909. The option was to expire on April 1, 1909 (less than three weeks hence).[5-3]

Studebaker Ignores the EMF Request to Parley

April passed without a response from Studebaker regarding the option. One reason for the silence may have been that, at the time this first proposal came to Studebaker, Frederick Fish had been on a six-week trip to the Pacific Coast. He returned after the 10-day option to buy had expired. On April 6, he met with Everitt and Flanders, but evidently nothing was resolved between the two companies. However, several Studebaker officers intimated that they would be willing to purchase a substantial amount of stock in EMF and would give the stockholders from whom they made the purchase a substantial profit over their original investment. Nevertheless, Fish received a strong impression that Everitt had every intention of terminating the EMF sales contract with Studebaker, regardless of any negotiations.

Two days after the meeting, Fish wrote a letter to Flanders in which, if one reads between the lines, there is the sense that he seemed to express surprise, astonishment, and disappointment over the fact that EMF could harbor such ill feeling about their sales contract. Nor could Fish understand what all the fuss was about, because the two companies had reached a good-faith agreement the previous August on the role of Studebaker in the sale of EMF 30 automobiles.[5-4] That role, Fish reminded Flanders, was that Studebaker would become the sole distributor of all EMF cars beginning September 1, 1909.

It would seem that both parties were in error in how they interpreted the loose language of the original agreement. The Fish version would mean the eventual elimination of Metzger and his sales department. Considering Metzger's sales talents and his friendship with each member of the EMF board, it would seem unlikely that the EMF officers would have backed an agreement in which Metzger and his sales force would be knocked out of the picture at the outset of the next model year. At the same time, Fish's actions easily could be interpreted as an attempt to maneuver EMF into the same position by which Fish had cowed Garford into a merger.

Fish took great pains to remind Flanders that Studebaker had made a substantial investment in the prospect of taking over the sales of all EMF 30 vehicles come September 1. Fish lamented,

> I cannot tell you what our losses are going to be on account of failure to make deliveries and on account of the car not being right in construction in several particulars.

Then he added in a more threatening tone (the italics are mine),[5-5]

> There is another serious question in regard to this, and that is, how can you possibly do such a thing [terminate the sales contract] in view of our expenditure of time and money, without feeling that *it is necessary for you to make a recompense to us for that time and money.*

The Glidden Tour Intrudes

In the midst of the turmoil involving Studebaker, Metzger had not been sitting idle. As sales director, he plucked the rather juicy plum of having the EMF 30 designated as the Pathfinder for the 1909 Glidden Tour.

The Glidden Tour was one of the most prestigious automobile events of the year. In essence, it was an endurance race for automobiles in which penalty points were awarded at the end of each day if an average speed of 20 miles per hour were not maintained or if repairs had to be made because of breakdowns on the road. At the end of the race, the driver whose make had the fewest number of penalty points received a trophy for his efforts.

The 1909 version would be the fifth and longest of the Glidden Tour series, extending approximately 2,636 miles from Detroit through Chicago, Minneapolis, Omaha, and Denver before finally winding down in Kansas City—after 15 days of travel. The roads, or rather the lack of decent roads, that had to be traversed would be more appropriate for modern four-wheel drive utility vehicles than automobiles of the 1909 era. For example, the initial route began on 10 miles of brick road leading out of Detroit, then shifted to deep sand for most of Michigan. The sand was so deep that drivers found themselves bobbing and weaving along as if they were shooting a series of endless rapids. So great was the dust cloud raised behind any vehicle that a following car had to trail at least 50 yards behind the other for the driver to see where he was going.

The race did not actually begin until July 12, but by virtue of being selected as Pathfinder, the EMF 30 had to set out on the road three months earlier, on April 12 with tour officials to measure, map, and photograph the route in time for publishing and releasing this information to the contestants. In one sense, the Pathfinder vehicle had a more difficult task than running in the race itself because it would be traveling on roads made even more treacherous than usual because of the spring thaw. When F.B. Hower of the American Automobile Association, who was Tour Chairman, was told that the Pathfinder would be an EMF 30, he asked if EMF did not have a bigger, stronger, and more powerful car because the trip would be extremely rugged. Hower warned that if the EMF 30 broke down en route, he would have no choice but to acquire another make. Aware of what Hower had said about the EMF 30, Dai Lewis, who would act as official Pathfinder, arrived at the EMF factory on Saturday, April 10, expecting to receive a specially prepared machine. Instead, he was taken onto the factory floor and, to his astonishment, told to pick any one of the 40 vehicles that had just been assembled. Lewis and his driver, a man named Meinzinger, then spent the entire Sunday driving over the worst roads they could find to see if the EMF 30 would be up to the task.

On Monday morning, April 12, Lewis, Meinzinger, and the EMF 30 huddled in downtown Detroit and, at the sharp report of a signal cannon triggered by the mayor, bolted out toward the open road. It did not prove to be an auspicious beginning. By nightfall, the participants realized that conventional EMF 30 springs were not adequate for carrying the unexpected extra 700 pounds of gear necessary for the long trip. To be more precise, the frame was bottoming on the rear axle. A wire was sent to the factory, and by noon the following day, the EMF 30 was back on the road with a new set of rear springs that had been fitted with two extra leaves.

The only other technical incident was the necessity to clean the carburetor in Kalamazoo. It was never adjusted thereafter during the entire 2,800 miles and 38 days of driving. Fuel usage was logged meticulously, 320 gallons having been used over the life of the trip. An indication of the state of the roads can be gathered from the fact that the EMF 30 had to travel 451 miles in low gear, 688 miles in intermediate gear, and 1,698 miles in high gear. Overall gas mileage was 8.82 gallons per mile, but the average in high gear was 14.18 miles per gallon.

Pathfinding in an EMF 30 for the Glidden Tour of 1909. (Courtesy of the Detroit Public Library, National Automotive History Collection)

Talk about road conditions! Out of curiosity, after slogging hub deep through gumbo mud for several hours and arriving in Mankato, Minnesota, the crew of the EMF 30 weighed their car and discovered that they had added 1,100 pounds to it—all mud. The early fears that the American Automobile Association officials had over the durability of the EMF 30 were dispelled. The vehicle performed as well as or better than the heavy, expensive machines that had been used for Pathfinder purposes on previous tours. The EMF 30 finally arrived in Kansas City with nothing having broken, nothing having failed, and nothing ever having bent. In short, the EMF 30 performed as if it had been driven sedately over city streets for the entire trip. After the car was washed, it ran as if it were quite ready to set out on the entire ordeal again.[5-6] (Later, as preparations for the actual Glidden Tour in July came to a head, three EMF 30's that had been entered were scratched from official participation on the tour. No reasons were made public. However, the vehicles did accompany the tour group over the entire trip, carrying members of the press who were reporting on its daily progress. A fourth EMF 30 acted as the lead or "pilot" car, carrying tour officials.[5-7] The EMF 30's performed their tasks quite well, but this time without any of the notoriety that resulted from the April Pathfinder trip.

For the 1909 Glidden Tour, which began in Detroit, cars lined up in Cadillac Square and started at intervals from in front of the Hotel Pontchartrain. (Courtesy of the Detroit Public Library, National Automotive History Collection)

Be that as it may, as the designated EMF 30 Pathfinder vehicle was slogging its way through sand and mud during the month of April, the company headquarters on the corner of Piquette Avenue and Brush Street was the scene of a very different kind of excitement.

Flanders Finally Responds to Everitt's Accusations

Walter Flanders found himself being attacked from two directions. On the one hand, the president of his own company accused him of becoming so friendly with Studebaker that he was neglecting his production duties. On the other hand, the erstwhile head of Studebaker was upset with him because he had not stopped Everitt's attempts to derail their sales contract.

Flanders made his initial feelings known—not to Fish, but to Everitt. On April 12, the day that the EMF 30 Pathfinder vehicle left downtown Detroit with great fanfare to scope out the Glidden Tour, Flanders fired off an angry letter to the EMF board. One can sense the anger in his words.[5-8]

> The only available cash which the directors of this company assured the general manager that he could use, after absorbing the Wayne Automobile Company and the Northern Automobile Company, was a hundred thousand dollars as a loan from the stockholders of the Northern Automobile Company and $200,000 as a line of credit at the banks. With this sum of $300,000, your general manager was obliged to equip and complete a factory for the production of these cars, and in addition to equipping buildings already available, it was found necessary to erect another building, practically doubling the original floor space and the machinery equipment, together with jigs, tools, and fixtures for the production of the car, representing an expenditure as our ledger shows, of $250,000.

> Why should the directors of the E.M.F. Company have any ground for believing that this company—with its limited working capital, with a factory building on its hands, without any equipment, and obliged to erect practically a duplicate factory before complete cars could be manufactured—could produce as many automobiles per day, or per month, as the Cadillac Company could with a completely equipped factory and a line of credit of $600,000 which that Company actually used.

These words were in direct rebuttal to Everitt's comments that Flanders had been neglecting his duties as general manager.

Furthermore, Flanders pointed out that as of the date of his letter, EMF had shipped 1,060 cars despite the previously mentioned obstacles, with 1,000 being the number he had predicted in August as the maximum he could attain over the next six months. Their profits would allow the fledgling company to weather the storm of financial adversity that all new companies must face. Flanders wondered what gave impetus to the directors' assumption that he could build two to three times that number of vehicles within the same time frame. The fact that EMF had received that many orders did not mean EMF could immediately build them. The two did not go together.

Moreover, Flanders was proud that he not only had accomplished everything he had set out to do on the limited budget given to him, but that he had drawn on only $25,000 of the $200,000 line of credit from the bank in doing so. Flanders attributed much of this success to the fact that, as a former, well-known machine salesman, he was able to equip the EMF factory with machinery received on credit from his friends.

Flanders Clears Up Questions on Studebaker Contract Terms

Flanders now explained the sales contract situation with Studebaker in very precise terms. He said that the sales contract was legally in force covering the period August 5, 1908, until September 1, 1909, after which it no longer was valid because of two technicalities: (1) it did not precisely specify the terms under which Studebaker would take over new sales regions from EMF salespeople, and (2) it did not fix either the time or the duration of the contract after September 1, 1909. However, Flanders added that the contract was signed in good faith by both parties initially, with the expectation that Studebaker would continue to handle the sales of EMF products following the latter date. Therefore, it would be prudent to resolve the matter fairly and on amicable terms with them.

Closer to home, Flanders outlined the harsh financial future that EMF might face if it terminated its relationship with Studebaker. He estimated that there would be the loss of approximately 5,800 units in sales between the date of his letter and March 1, 1909. More importantly, EMF would forego a profit of $540,000.

The old adage that "money talks" could never be put to better advantage as Flanders outlined other financial considerations to the board. Was the board aware that, as testified by Henry Joy before Congress, eight of ten automobile companies go bankrupt between the months of November and the following February? Why? Simply because the only people buying cars during those months are those who live in less populous, warm climates (i.e., the regions that Studebaker agreed to handle). Flanders then said that if EMF should find itself with as few as 900 to 1,200 unsold vehicles on its hands over the next six months or year, the loss of profit from the sale of those vehicles would force EMF into bankruptcy.

Then came the clincher[5-9]:

> Are we prepared to spend at least $100,000 in conducting a sales campaign in this territory, already covered by the Studebaker company?

Flanders urged the board to come to terms with Studebaker.

A New Sales Contract Is Proposed to Studebaker

Flanders' long exhortation in favor of retaining Studebaker did not fall on deaf ears. Despite Everitt's determination to rid EMF of the Studebaker presence, board members decided to make a new proposal regarding that aspect of the existing sales agreement covering the period after September 1, 1909.

On Wednesday, April 21, the board met in a special session and hammered out new words for the sales contract. It was based on Flanders' projected production estimates for the period September 1, 1909 to the end of August 1910. In the proposal, EMF offered to build 3,600 automobiles for purchase by Studebaker during that 12-month period to sell out of its agencies. The proposal further included an option by which Studebaker could buy 1,200 additional cars during this period. The price for each EMF 30 received by a Studebaker dealer for retail sale would be $1,250, payable on delivery minus 28 percent. This would bring Studebaker's actual unit cost to $900. Assuming that the dealer would sell the EMF 30 for $1,250 (i.e., its advertised price), Studebaker would realize a handsome $350 profit per automobile, or a total of $1.25 million during the life of the contract. Of course, there also was a list of optional items such as tops or windshields. These items would increase the cost of the car to Studebaker, but that cost would easily be recovered because the customer would have requested the optional items. The EMF directors added that under no circumstances was Studebaker to attempt to sell an EMF 30 for more than $1,250. Moreover, Studebaker was to retain a sufficient supply of repair parts in its branch houses to ensure that customers received quick service. These parts would be shipped each month, and Studebaker would be billed for them immediately (less 20 percent off list price).

The critical part of the new proposal was that it restricted Studebaker sales outlets of EMF vehicles to all states west of the Mississippi River except Minnesota and South Dakota, plus the territories of Canada and Mexico, and the Southern states of Alabama, Georgia, Florida, Mississippi, North and South Carolina, and Tennessee. Under this scenario, the more populous states in the Midwest and on the East Coast would continue to be the direct responsibility of Bill Metzger and the EMF sales staff. A caveat, if you could call it that, was that any automobile sold by Studebaker within its sales territories would be called the "E-M-F-Thirty-Studebaker."

All the more audacious was the final point in the proposal. It stated that EMF was in the process of developing a low-priced car to sell at $500 to $650 and intended to make 6,000 of these cars during the coming year. (This was Flanders' dream as well—to compete directly with Ford.) If these cars were built during the terms of the contract, Studebaker would be required to purchase the entire production run for resale at a price still to be determined![5-10]

The vote of the five EMF board members in favor of forwarding the proposal to Studebaker was unanimous, and the board moved that Studebaker be given 10 days to accept or reject the proposal. Walter Flanders seconded the motion, which was all the more surprising because he had taken it upon himself to conduct all previous transactions between Frederick Fish and Studebaker, bypassing EMF president Barney Everitt. Moreover, Flanders knew that Frederick Fish probably could not live with the proposal because the latter had made it quite evident that he wanted Studebaker to have complete control of all EMF sales come September 1909. The proposal kept Everitt happy because it would place a leash on the future ambitions of Studebaker that could not be avoided if Studebaker expected to continue its relationship with EMF. That leash would be secured only if Studebaker signed on board, however. Because this was highly unlikely, Everitt finally would be rid of Studebaker.

How wrong Everitt would be. What occurred eight days after the proposal was dispatched completely shocked the automotive world.

Everitt and Metzger Are Out; Studebaker Is In

During the week following the dispatch of the new sales proposal, Everitt discovered that he was the one out of sync with most of the EMF board, especially Flanders. This soon became evident when Studebaker had made a counter-proposal that was too good to be rejected.

In an unprecedented move, Studebaker offered to buy out the stock holdings of Everitt and Metzger. Their combined holdings represented more than one-third of the shares outstanding, according to the first EMF annual report to the state of Michigan. Everitt owned 3,500 shares, and Metzger owned 4,500 shares. The two also had made further purchases prior to the sale, as recorded in the stock certificates issued after the annual report was made. Therefore, their total investment in EMF was approximately $180,358 by the end of April 1909. The purchase price tendered to them has been reported variously as $1 million, or $800,000, or $362,000. In any event, the Studebaker offer either doubled or tripled the amount of their original investment. Because the price of EMF stock had not changed much during the intervening months, the two EMF officers realized a substantial profit from their year-long participation as EMF officers and directors. To realize the enormity of it, consider that the average worker in an EMF plant would have to work more than 240 years to realize even the low end of these numbers at the prevailing wage rate of that time.

The move by Studebaker was what finally brought peace to the EMF front office. Everitt and Metzger sold out. In fact, the sale brought more than peace. Now Flanders could rely on the deep pockets of Studebaker as stockholders to help build a strong automobile company. Unfortunately, Flanders also would discover that the fears of Everitt and Metzger about the heavy-handed presence of Studebaker would become more fact than fiction.

Chapter Six

Flanders Expands EMF, Declares War on Studebaker

With the departure of Everitt and Metzger from EMF, it was only a matter of time before the egos of Fish and Flanders eventually would collide. Frederick Fish of Studebaker and Walter Flanders of EMF had come together with the intent of using each other to achieve their respective goals. For Fish, that goal was to have Studebaker become a power within the automobile industry, without having to go through the effort of building a factory from the ground up. For Flanders, it was to raise EMF into an automotive giant that would rival Ford Motor Company.

Their respective worlds began peacefully enough, however. First, there was the need to replace Everitt and Metzger.

On April 29, 1909, an emergency meeting of the EMF board members took place. Only three were present: James Book, Charles Palms, and Walter Flanders. The emergency? To accept the formal resignations of Barney Everitt and Bill Metzger as directors and officers of EMF, and to release them and the chief engineer, William Kelly, from their contracts as well as from any liability the company may have accrued during their stay. John Beaumont, the lawyer representing Everitt and Metzger, presented a letter to the board, stating that they had accepted the proposal of Clement Studebaker, Jr. and Frederick Fish to buy their EMF stock. However, the sale was contingent on the EMF board agreeing to release Everitt and Metzger from any contractual obligations they had with the company, as well as any liability toward four outstanding bonds still due with two Detroit banks that the EMF directors previously had signed. They also asked to be paid their salaries up to October 1, 1909, plus five percent of the EMF net earnings realized to that date. The letter also made it quite clear that Everitt and Metzger were leaving EMF because they were "out of harmony with what seems to be the desire of this board in respect to the proposed future relations between this Company and the Studebaker Automobile Company."[6-1] (The Studebaker Automobile Company was a separate company set up by Studebaker Brothers to purchase and market the EMF 30 cars.)

The three remaining board members immediately passed a resolution, accepting the terms tendered in the Metzger and Everitt letter. They also agreed to Kelly's resignation; however, in recognition for his efforts on behalf of EMF, the board awarded Kelly a cash equivalent of the stock that he would have received if he had remained with the company for the entire three years of his contract. The latter money would be paid in three installments of $11,666.67, their due dates being October 1 of 1909, 1910, and 1911.

That business being finished, Frederick Fish and Clement Studebaker, Jr. were voted onto the EMF board. Walter Flanders then was elected president in place of Everitt. Charles Palms took Metzger's position as secretary.

No sooner had Fish taken his seat on the board, then he quickly made his presence felt. First, he introduced a new proposal for handling EMF sales, written by Hayden Eames, Studebaker general manager. It was a rewrite of the existing sales agreement between the two companies. (Had both Everitt and Metzger still been around, they probably would have voted against it.)

In essence, Eames recommended that the existing sales contract should remain in force until the end of August 1909. Beginning September 1, 1909, however, and continuing over the next three years, new terms would be included, stating that the Studebaker Automobile Company would be the sole distributor of every automobile EMF could manufacture. Furthermore, beginning with the 1909 models, all EMF 30 cars would be badged as "Studebaker E.M.F. 30" models.

Other than those two changes, Eames left the original contract intact. EMF would continue to ship its cars to Studebaker at a cost of $900 per unit, payment to be made on their delivery at the point of sale. Repair parts made by

EMF and sold to Studebaker would be priced at 160 percent over cost, minus 20 percent. EMF would be responsible for guaranteeing each car and taking care of any defects that might arise. For its part, Studebaker would not sell cars that were made by any other manufacturers priced between $850 and $1,750.[6-2]

Fish asked the EMF board to terminate the existing sales contract between EMF and Studebaker in favor of the Eames proposal. His resolution passed unanimously.

Fish also recommended that Walter Flanders be awarded 12,500 shares of fully paid EMF stock to compensate for the commissions he would have received under his previous contract as general manager. This resolution also was adopted unanimously.

Fish estimated that EMF had saved approximately $750,000 by not having to pay all the stock options due to Everitt, Metzger, and Kelly under their former contracts. This savings and others that came from the new sales contract with Studebaker would only enrich the stockholders down the road.[6-3]

With Walter Flanders now the titular head of EMF, someone was needed to act as general manager to handle the day-to-day operations of the company. No problem. The board offered Flanders a contract to also act as general manager until June 1, 1913, at a salary of $6,000 per year (approximately eight times that of the average EMF assembly worker). In addition, Flanders would receive a bonus of five percent of the net profits of EMF as of September 30, 1909. The board then offered Charles Palms a contract to act as EMF treasurer for four years at a salary of $5,000 per year.

The board met again the following morning at 11:00, at which time its members were notified that Flanders and Palms had signed their contracts and that a written acceptance of the new sales contract between EMF and Studebaker had been delivered to the latter company. The board then added Hayden Eames to its ranks as a director.

The past two days had seen a pronounced turning point in the life of the young company. In effect, with the adoption of the new sales contract and the nomination of new officers and directors, EMF had been reorganized.

By the end of summer 1909, the maneuverings of Frederick Fish to take ultimate control of EMF and merge it with Studebaker would become all the more obvious to Flanders as time passed, now that Flanders was president of EMF. In fact, one of his early moves was to squash just such an attempt by Fish to convince the board to agree to such a merger.

As president, Flanders had to be pleased with what EMF had accomplished thus far. As of April 1909, less than one year since it had been formed, the company had produced 7,500 automobiles, more than any other car maker in the country except for Ford and Buick for the entire 1909 model year. Yes, it was a most satisfying year.[6-4]

From Penny Pinching to Bold Expansion: The De Luxe Purchase

Backed by the deep pockets of Studebaker, Flanders now began to kindle the fire of his real dream—to add a new, low-priced model that would compete with the Ford Model T. So as not to take away from the time he was forced to spend building up EMF 30 production to levels that he found acceptable, Flanders decided to purchase another car company and use its manufacturing facilities as a base on which to build a low-priced EMF automobile.

In July 1909, Flanders made his move. The company he selected was the De Luxe Motor Car Company, located on a 15-acre site at Clark Street and Jefferson Avenue on the west side of Detroit, not far from the Cadillac complex that General Motors would build more than a decade later.

De Luxe was a transplant from Toledo. It had begun life in the factory that belonged to the Kirk Manufacturing Company, maker of the Yale car which Bill Metzger had helped sponsor. Kirk had quit car manufacturing in 1905 and left the Yale plant vacant. De Luxe took over the plant in 1906, then shortly thereafter left it to move into the recently built Clark Street factory in Detroit which it had acquired from C.H. Blomstrom, whose attempts to build a car called the Queen had failed.

For three years, De Luxe had tried to market a vehicle in the $5,000 range but had met with little success. One reason was that the car was so costly to build that no profit margin remained. For one example, De Luxe used the most expensive sets of ball bearings on the market.[6-5]

By 1909, De Luxe was ripe for a takeover. It had built fewer than 100 cars the previous year—far too few to keep a factory of its size open for very long.[6-6] Several parties already had been negotiating to buy the property when Flanders entered the bidding battle, made his proposal, and then delivered the money to conclude the purchase—all within the same day. For an estimated $800,000, Flanders took title to the De Luxe land, buildings, machinery, patents, drawings, tools, and fixtures. Also that same day, he dispatched 150 men from the EMF plant to begin the task of converting the De Luxe plant into one that would produce a second EMF model.

Shortly thereafter, Flanders publicly announced that the De Luxe property would become home for a new four-cylinder runabout on a 100-inch wheelbase that would be marketed as the Studebaker-Flanders 20. Expectations were that the new model would go on sale the first of January 1910, for approximately $750. This selling price alone was enough to force the industry to sit up and take notice, because the car would have been cheaper than all but the lowest-priced Model T projected for 1910. Moreover, the Flanders 20 would be built by Walter Flanders, the man instrumental in bringing the Model T into existence when he worked at Ford.

Flanders also made it known that he intended to build 25,000 Studebaker-Flanders 20 units during the 1910 model year. Again, this must have caused eyebrows to raise, inasmuch as no car company to date had built that many automobiles in a single year. (Ford and Buick were the only makes to have exceeded 10,000 units.) Furthermore, the 25,000 did not include the build of the well-accepted EMF 30 models. If things went as planned, Flanders was about to stand the auto industry on its ear.

Interest in the Studebaker-Flanders 20 mushroomed so quickly that Studebaker felt it necessary to begin advertising the vehicle six months before it would be offered for sale. One of its ads read[6-7]:

> Out of consideration for competitors, we intended to keep silent yet a while. Deliveries will not begin until January. Besides, we realized that the announcement of such a car at

such a price, and by Studebaker, is likely to have the effect of an explosive bomb on the market at this time. We had no desire to precipitate anything, but our hand was forced.

Clearly, although the Studebaker-Flanders 20 was the brainchild of Walter Flanders, Studebaker was sending a message through its ads that the car would be a Studebaker, albeit one built by EMF.

In a sense, there might be some merit in Studebaker's advertising direction. With the departure of William Kelly, EMF no longer had a chief engineer. Although Flanders may have been a production genius, he was not an engineer who could design the new Studebaker-Flanders 20. To remedy this lack of expertise, Flanders hired James Heaslet, a self-taught engineer who had migrated around various small car companies in the area and was then working for Studebaker.

Birth of the Studebaker-Flanders 20

Self-taught or not, Heaslet had completed a design for a new four-cylinder car that met Flanders' approval by the time of the De Luxe purchase. Moreover, the new car incorporated a rather ingenious solution for alleviating the rigors of engine repair. Heaslet mounted the engine, magneto, carburetor, radiator, pump, steering gear, and dash panel on a special sub-frame made of two parallel steel tubes. The sub-frame in turn was bolted to cross-members bridging the frame. By simply removing four bolts, the entire sub-assembly could be lifted out of the body in about five minutes, so that it could be worked on in the open.

The ever eloquent E. LeRoy Pelletier made the sub-frame one of main attractions of the car later when advertising the Studebaker-Flanders 20.[6-8]

> We expect this feature to revolutionize present garage practice which necessitates laying up the car for days at a time while some minor repair is being made. In case of any repair or replacement in a Flanders "20," however serious or simple, the easiest way is to replace the entire unit, send the owner away rejoicing with his car, and then, when time best suits

and with parts most accessible, make the necessary repair at a minimum of time and expense. The original unit may later be replaced in the car—or if the condition, as to wear, of the two units are about the same, the change need not be made—the owner simply charged for time and material in making his unit good.

The Studebaker ad also provided the rationale for predicting a production run of 25,000.

This quantity was necessary. It would be impossible to produce a car of this size and quality at the price if made in smaller quantities. The tremendous "overhead" expense of equipment and distribution would, if saddled onto a lesser number of cars, make it necessary to add 25 to 50 percent in the price. We cannot build a much better car than others do for the money in lots of 3,000 to 5,000. But by distributing the overhead over 25,000 cars, we have been able to set the price at $750.

One question immediately comes to mind: If EMF had been able to build only half the 25,000 units, would the price of the Studebaker-Flanders 20 have risen perceptibly?

Those skeptical of the optimism of Studebaker and EMF about the future number of cars they intended to manufacture may have had cause to change their minds when Flanders announced that work soon would begin on enlarging the Piquette production facilities. The existing plant would be increased to four stories, and new buildings of that same height would be constructed.[6-9]

Four Supplier Plants Added to the EMF Empire

These were only the beginning of EMF efforts to expand its operations. Similar to Henry Ford, Walter Flanders also felt that his plant must manufacture as many of its own parts as possible. To this end, one week after the De Luxe purchase, Flanders added the Western Malleable Steel & Forge Company and the Monroe Body Company to his budding empire. Both companies already made parts for EMF. About Western Malleable Steel & Forge Company, Flanders was quoted as saying,[6-10]

> I believe that we are the first concern to own its own forging plant.

The Western plant adjoined the De Luxe factory and cost Flanders $300,000. It initially was founded to build malleable steel castings; however, the owner had discerned a trend in the auto business toward drop forgings and had gone into their production in a big way. Some of the principal parts of an automobile engine such as crankshafts, front axles, spindles, gear blanks, and connecting rods were coming from the Western shop.[6-11]

Owning the Monroe plant in Pontiac, a few miles north of Detroit, meant that EMF now could make its own bodies. The cost for Flanders to purchase the Monroe Body Company was $200,000.

To these acquisitions was added the Sanitary Steel Stamping Company, a bathtub company that was converted into a plant to make automobile stampings. It featured a huge (for that era) press capable of 2,500 pounds pressing power. The use of pressed steel by the auto industry was only in its infancy, but Flanders was attracted to its possibilities for reducing vehicle weight, especially in lieu of heavy castings. The bathtub plant represented the first pressed steel plant of any consequence in use for automobile production on a daily basis. In time, fenders, hoods, running boards, body panels, and fuel tanks issued forth from this facility. Small parts such as control lever brackets, hub caps, radiator caps, rear axle housings, and running board brackets also became a staple part of its production capabilities.[6-12]

More plants were targeted for purchase. To make its own crankshafts, EMF purchased the Auto Crank Shaft Company. The former Northern plant in Port Huron, Michigan, also was put to work making bevel gears and rear axles, and on empty land next door to the De Luxe plant, a brass foundry and iron works were constructed. What made this expansion all the more surprising was that it was paid for entirely by the profits that EMF had realized during its first year of operation.[6-13] At least, this is what Flanders reported at the October 12, 1909 meeting of the board of directors.

EMF and the 1909 Glidden Tour

As if these activities were not enough to keep EMF in the news, E. LeRoy Pelletier, who had just been appointed assistant general manager of the Studebaker Automobile Company, was up to his publicity-gathering tricks. During the weekend before the Monday, July 12 start of the Glidden Tour from downtown Detroit, Pelletier made it a point to regale the press that had gathered for the event rather than cozy up to Glidden Tour personnel and Detroit city officials. On Friday, he treated reporters to a frog leg dinner in the grille room of the Lighthouse Inn. He did the same again the following evening. On the second occasion, the 34 visiting newspapermen rose to toast Pelletier, who thanked his guests (no doubt tongue-in-cheek) for enabling him to be promoted to his new position.

As the summer of 1909 came to a close, it seemed as if anything Flanders attempted to do was signaled with success. Most importantly, EMF was reported to have had a perfectly marvelous first year of existence, bringing in profits of approximately $1.4 million on an investment of only $200,000.[6-14]

Flanders Sets EMF Production Targets for the Coming Year

As the EMF directors gathered for their October 12, 1909 meeting, Flanders clarified any misconceptions about the past output of the company. He reported that thus far, the company had built and sold 6,074 EMF 30 cars. Moreover, Flanders added,

> ...business conditions demand that we manufacture during the next twelve months from 25,000 to 30,000 cars, of which from 12,000 to 15,000 will be the E-M-F "30" model and from 14,000 to 16,000 will be of the Studebaker-Flanders "20" model.

Flanders offered a monthly production schedule, extending from September 1909 through August 1910, by which this number would be achieved. From the tone of his report, it appears that Flanders fully expected to reach this extraordinary (for that time) volume of manufacture. The schedule called for the production of the Studebaker-Flanders 20 to be 800 units during its first build month, January 1910.

It is a wonder that the other members of the board did not flinch at such an outrageous production schedule. Such an output would place EMF on the same level as the Ford Motor Company and place it on the path of becoming the Number One or Two auto maker in the nation, if not the world. Evidently, everyone had confidence in Flanders' ability to achieve such an outlandish goal. Their confidence demonstrates the effect that Flanders' dynamic, larger-than-life personality had on those around him.

How to Build a Loyal Workforce

Indeed, Flanders did have the EMF factories running to a schedule of 1,000 cars per month at the beginning of September 1909. Flanders attributed much of this to the working relationship he had with his men—a relationship based on a bonus system that extended downward to everyone in the EMF workforce, even the office boys and stenographers. By virtue of this bonus system, Flanders honestly believed that his workers would make every effort to achieve the goals he set for them. Production numbers of the previous day were posted in the factory the next morning. If output slipped below what was expected for that day, a general inquiry was made. In five minutes, everyone would know what precipitated the delay and where it occurred.[6-15]

One example of the rapport Flanders had with his men was an offer he made to them on October 12, the same day as the October EMF board meeting. It was the opening day of the 1909 World Series in which the Detroit Tigers were playing against the Pittsburgh Pirates. The game was at Bennett Field, an easy streetcar ride from the EMF factory. Flanders left word that if the factory could reach its scheduled 50 units of production by noon, everyone could have a half-day holiday. Those who wished could attend the game. (The Tigers won the game, 5-0, but lost the series.) Flanders' offer stood for each of the four game days scheduled to be played in Detroit. The scheduled output indeed was reached on each of those game days.[6-16]

Flanders and EMF were on a roll. There was no question about it. Studebaker outlets had taken over the sales of all EMF automobiles, per the new contract, as of September 1909. The enormous leverage they provided by being positioned across the United States and in other countries gave Flanders confidence in his belief that his production goals were not unrealistic. He felt that Studebaker was perfectly capable of selling each of the 1,000 EMF 30's he would produce monthly between September and December 1909. These sales would provide EMF with a gross income of $3.6 million and enable the company not only to pay all its bills for this period but to further expand operations as well. Long forgotten were the lean, startup days of the preceding year.

Flanders was EMF. He was its driving force. What success the company achieved was due to his efforts and those of Studebaker. In recognition of what he had thus far achieved, the EMF board in November unanimously voted to increase his salary from $6,000 to $25,000 per year, retroactive to the beginning of September.

This was the last happy EMF board meeting for some time.

Flanders Denounces Studebaker, Takes Strong Action

Near the end of November 1909, Walter Flanders began to realize that perhaps Everitt and Metzger had been correct in their evaluation of Studebaker's intent in aiding EMF. True, Studebaker had purchased the required 1,000 EMF 30's produced in the month of September. However, during October and November, the Studebaker purchases had dropped off at an alarming rate. Rather than selling every car that EMF could produce, as Flanders had expected and the contract read, Studebaker accepted the delivery of 819 EMF 30's in October and only 416 in November. Flanders sensed that Studebaker's stall contained all the elements of a squeeze play to precipitate EMF into a position whereby it could be forced into bankruptcy, with Fish waiting in the wings to pick up the pieces at bargain prices.

The Studebaker representatives on the EMF board (i.e., Frederick Fish, Hayden Eames, and Clement Studebaker, Jr.) were well aware that Flanders had incurred serious expenses to maintain a production level commensurate with his forecast. These expenditures amounted to $1.024 million in September, $1.031 million in October, and approximately $920,000 in November.[6-17] In fact, the three had voted with the other EMF board members to place a hold on further cash stock dividends until all plants, buildings, machinery, and equipment had been paid for and an adequate reserve fund had been put in place. To pay for this outflow of funds, it was imperative that 1,000 EMF 30's be sold during each of these three months, so that their revenues would offset accrued liabilities. The three EMF board members from Studebaker also were quite aware that under the terms of the sales contract between the two companies, EMF could not sell its product through any other agency. Thus, if Studebaker was "unable" to meet its sales quota and EMF could not pay its bills... Well, let nature take its course. Or so Fish may have thought.

Flanders Goes on the Offense

Walter Flanders' initial reaction to what he suspected of Studebaker was mild. After informing William Barbour, James Book, and Charles Palms (the non-Studebaker board members) of his suspicions, he wrote the following letter to Studebaker:

> Detroit, Mich., November 23rd, 1909.
>
> The Studebaker Automobile Company,
> South Bend, Indiana.
>
> Gentlemen:
>
> As you are well aware, the Studebaker Automobile Company failed to take 181 E.M.F. cars of the 1,000 specified for the month of October last, and thus far we have only received shipping directions for 397 of the 1,000 E.M.F. "30" cars specified for the month of November.
>
> The E-M-F Company has been, at all times, during the months of October and November, ready to ship these cars to you according to the specified schedule and cannot permit shipments to fall and remain below the schedule. We, therefore, request that you send to us at once shipping directions for 181 cars, being the balance of the number specified in the

schedule for October and also shipping directions for sufficient cars for the month of November to make up the full 1,000 in number specified by you.

Yours truly,

EVERITT-METZGER-FLANDERS COMPANY
W.E. Flanders,
President and General Manager

One week passed without any response from Studebaker. Now beginning to lose his patience, Flanders fired off a telegram on November 30, 1909, in which he said,[6-18]

> Have not yet received answer to letter of November twenty-third. We insist that you immediately give us shipping instructions for seven hundred forty-eight cars being balance of full number of two thousand cars specified by you to be taken and paid for in the months of October and November 1909.

December arrived. One day passed, and then another. Finally, on Friday, December 3, Walter Flanders, James Book, and Charles Palms met with Frederick Fish and Hayden Eames. The discussion was heated but inconclusive. Fish was adamant that "a matter of such importance as this must be passed upon by your Board of Directors."[6-19]

However, time was running out. New EMF 30 automobiles were gathering moss at the factory, taking up space, running up insurance costs, and bringing no income to offset material and production expenses. The patient Flanders dispatched another letter to Studebaker on December 6, 1909. In it, he reiterated the September 1909 through August 1910 EMF schedule of deliveries to Studebaker, which amounted to 15,200 cars.

What truly irritated Flanders was that Studebaker thus far had ordered only 222 cars for the month of December. Because the latter already was well arrears in its orders for September through November, things did not bode well for the solvency of EMF. Flanders reminded Studebaker (as if Fish were not already aware) of EMF expenditures, and that the unordered automobiles denied EMF almost three-quarters of a million dollars in profit to offset expenses. Again, Flanders asked whether Studebaker expected to honor its commitment to "take and pay for the automobiles ordered by it from September 1, 1909, to September 1, 1910, in accordance with the monthly schedule of deliveries agreed upon."[6-20] Flanders requested an immediate answer.

Again, none was forthcoming.

Finally, on December 9, after caucusing with the three sympathetic EMF directors, Flanders made the ultimate move. His letter of December 9 severed all connections between EMF and the Studebaker Automobile Company[6-21]:

> You are hereby notified that we have elected to and do now treat as rescinded and annulled all contracts and agreements made and entered into between us whereby it was stipulated and agreed that the Studebaker Automobile Company should act as the sole distributor of the products of this Company.

Flanders listed four reasons for the rescission: (1) Studebaker was forcing its dealers to accept an unfair and unacceptable discount for selling EMF 30 automobiles; (2) the Studebaker dealers were required to sell the low-selling, expensive Studebaker-Garford gasoline-powered cars and their equally low-selling, expensive electrics to receive an EMF franchise; (3) Studebaker had advertised there would be no change in the 1910 EMF 30 models although the company knew that EMF had improved its suspension system and had replaced the front axle and clutch with new designs; and (4) their ads were worded to give the impression that Studebaker owned a controlling interest in EMF.

The last paragraph of Flanders' letter issued the EMF ultimatum: As of December 9, 1909, the company would sell the EMF 30 and Flanders 20 automobiles independently of Studebaker and would immediately negotiate with dealers and agents across the United States to arrange for the sale of EMF products.[6-22]

Flanders Dumps the Studebaker Automobile Company

When Flanders said "immediately," he meant "immediately." Two days later, on Saturday, December 11—while a shocked Frederick Fish, Hayden Eames, and Clement Studebaker, Jr., were digesting the contents of Flanders' letter of December 9—full-page ads began appearing in major newspapers, announcing that EMF had terminated its sales contract with Studebaker. Included was a reproduction of Flanders' letter to Studebaker, informing the latter of the rescission.[6-23]

Equally important was an ad worked out between Flanders and E. LeRoy Pelletier, which read,

> The E-M-F Company is ready to close with a representative in thirty-five American cities to handle its product direct, instead of through the sales organization which formerly acted as its distributor.

The ad was printed in half-page form. Its copy was telegraphed to the two leading newspapers in the 35 largest cities in the country. Pelletier had resigned as advertising manager for Studebaker and joined Flanders at EMF in the same post as soon as the break had become official.

Publication of the ad had an immediate effect. For example, on the following day, six men with substantial financial support were alleged to have boarded a train in San Francisco for the sole purpose of heading to Detroit to secure one of the agencies. They purchased 300 EMF 30's, paid for 50 of the vehicles in cash, and provided shipping instructions for the remainder. Within 10 days, EMF sold the 1,200 cars that Studebaker had not accepted, and 400 more were on order. The 1,200 sales brought EMF a million dollars.[6-24]

Another source reported that incoming trains were bringing in dealers daily. They were perfectly willing to wait in line to be interviewed as prospective sales agents. Sometimes, two or three were vying for the same territory. San Francisco produced seven candidates, each with certified checks for future purchases, contesting for the West Coast franchise.[6-25]

An EMF 30 in typical road race trim, circa 1910. (Courtesy of the Detroit Public Library, National Automotive History Collection)

As part of a publicity campaign, EMF 30's are at the starting line of a race in Savannah circa 1910. (Courtesy of the Detroit Public Library, National Automotive History Collection)

Endurance runs were part of a publicity campaign to promote EMF. This EMF 30 was embarking on an endurance run from Detroit to Rutland, Vermont, in 1911. Rutland was the birthplace of Walter Flanders. (Courtesy of the Detroit Public Library, National Automotive History Collection)

Chapter Seven

Crisis or Comedy? Studebaker Sues EMF

One cannot help but believe that Frederick Fish grossly underestimated the bulldog tenacity and popularity of Walter Flanders, who was a man not only of great strength but of great willpower also. Daring to pit the young EMF enterprise against the power and wealth of the leading wagon manufacturer in the United States was not a task for the faint-hearted.

No doubt reports of the large number of dealers seeking EMF sales franchises in the matter of a few days may have taken its toll. Conceivably, if Studebaker did not take quick action, its dealer body could find itself stripped of all but Garford's large, slow-selling luxury cars and electric models. The EMF 30, on the other hand, was selling well, and Studebaker made more money on it than did EMF itself.

In retrospect, Studebaker's game plan for taking over EMF, the same plan by which it had captured Garford, was in trouble. New measures were necessary, and they would be played out in the courts over the next three months as Studebaker desperately fought a fierce rear guard action to retain its sales agreement with EMF. At times, the ongoing struggle would take on the appearance of high comedy.

Studebaker Strikes the First Blow

On Monday, December 13, the Studebaker forces finally sprang to life. As soon as the business day unfolded, the Studebaker legal team filed a bill of complaint against EMF in the U.S. Circuit Court, Eastern District of Michigan, Southern Division. It was represented by the imposing legal talents of John Miller of Chicago, plus General Henry Duffield and Otto Kirchner of Detroit.

The complaint, first and foremost, asserted that the Flanders letter dissolving the sales agreement with Studebaker had not been formally blessed by the EMF board of directors; therefore, it carried no legal weight. This being the case, it was fraudulent.

Furthermore, unless the court intervened, Studebaker believed that Flanders soon would call a meeting of the four EMF directors supporting his actions for the sole purpose of passing a resolution to terminate the contract. The opposing directors (Frederick Fish, Hayden Eames, and Clement Studebaker, Jr.), who represented one-third of the outstanding capital stock, thus would be left with no say in the matter.

The EMF newspaper advertisement reporting the cancellation of the Studebaker sales contract also came under fire. Studebaker again argued that the EMF board did not sanction the publication of such an ad. More importantly, unless the court squelched the ad, Studebaker feared it would lose all of "the trade and customers and good will" that the company deserved. This, in turn, would disrupt and destroy the existing Studebaker organization because the Studebaker sales agents no longer would receive any vehicles to fulfill their part of the sales contract.

Finally, the complaint denied all allegations made by EMF within the advertisement; in fact, it denied that Studebaker had ever agreed to purchase vehicles in accordance with the production timetable Flanders had set for the 1910 model year.

What Studebaker Asked of the Court Against EMF

The Studebaker complaint posed four requests to the court:

1. That the court issue a subpoena against the Everitt-Metzger-Flanders Company, Walter E. Flanders, Charles L. Palms, James Burgess Book, and William T. Barbour;

2. That EMF be required to continue meeting the terms of the contract with Studebaker;

3. That EMF be perpetually enjoined from selling, shipping, or contracting with anyone else for its automobiles or any other of its products other than Studebaker or from interfering with Studebaker agents from selling the products, and that EMF be especially restrained from continuing to advertise for dealers; and

4. That the court issue a preliminary injunction restraining EMF from further advertising for dealers.

Obviously, the quick and virtually overnight success of EMF in attracting independent dealers had thrown a scare into Frederick Fish and the Studebaker forces. They were determined to retaliate with all their resources. The gloves were off. After filing the bill of complaint, John Miller, the lead attorney for Studebaker, was quoted as saying[7-1]:

> The Studebaker Company had observed and has not broken its contract with the E-M-F Company. It has established throughout the company selling agencies to the number of several hundreds for the handling of the E-M-F car... The Studebaker Company has never failed to pay for cars and the E-M-F Company owes a large amount of money under the contract. The claim that the Studebaker Company has failed to take cars required by the contract is untrue. The E-M-F Company has not yet manufactured or had ready for delivery the number of cars it claims the Studebaker [Company] is in default of taking. The Studebaker Company has taken all the cars it is obligated to take, and stands ready to take and pay for all such cars in the future.

On Tuesday, the day after the bill of complaint was filed, Judge Henry Swan refused the request of Studebaker to issue an order to halt EMF either from publishing its ad or contracting with new dealers. What the judge did was to order Studebaker to provide EMF with a copy of the complaint by Wednesday (the next day), and he required EMF to respond to the complaint by Saturday, December 18. He then scheduled a hearing for Monday, December 20.

The judge's decision was neither what Studebaker had desired nor expected. EMF was not halted from selling cars on its own, nor from signing up new dealers. Yet, those efforts had to be stopped before matters became out of hand. Already Studebaker was hearing from its sales agents across the country, who found themselves in a pickle. Were they authorized to continue selling the EMF 30 or not? If not, what were they supposed to do with the EMF cars they had on hand?

Studebaker Files a Second Suit

To circumvent Judge Swan, the Studebaker lawyers took their case to the Cincinnati court as quickly as they could. There they went through much of the same motions as in Detroit, asking the U.S. Circuit Court for the Southern District to issue a temporary injunction restraining EMF from selling its own cars to anyone else but Studebaker. The suit was filed on behalf of Messrs. Fish, Studebaker, Jr., and Eames against the other directors of EMF, a slight change in the wording of the principals which eventually would haunt Studebaker.

The ploy worked. Judge Henry Severens, apparently unaware of what was taking place in Michigan, granted the injunction on Thursday, December 16. He scheduled a hearing for the following Wednesday, December 22, in Kalamazoo, Michigan, a city on the western side of Michigan, where he would be sitting that day. Flanders and the non-Studebaker EMF directors did not receive news of this injunction until late Friday, December 17. On Saturday, they also discovered that Studebaker had placed large ads in the local newspapers declaiming that EMF 30 and Studebaker-Flanders 20 cars would continue to be sold by the Studebaker Automobile Company as in the past.

EMF appeared to be unperturbed by all of this legal activity. All through that preceding week and into the next week, Flanders continued to keep production high, turning out 60 EMF 30's per day and placing them in storage from where they could easily be removed and offered for sale as soon as EMF received a favorable ruling from the courts. EMF lawyers also had spent the week preparing their answer to the original bill of complaint, the hearing for which was due in Detroit on Monday. Although the injunction from the Cincinnati court no doubt jarred EMF, coming out of the blue as it did, it did not alter its course of action.

When the EMF attorneys appeared before Judge Swan in Detroit on Monday December 20, as ordered, they fully expected the hearing on the Studebaker bill of complaint to take place. They were enormously disappointed. Judge Swan, who by now was cognizant of the temporary injunction served on EMF by the District Court in Cincinnati and of the hearing scheduled in Kalamazoo for the coming Wednesday, deftly stepped aside. He rescheduled his own hearing for Monday, December 27, despite furious protests from EMF lawyers, who argued that they were fully prepared to proceed with the case. The delay also meant that the temporary injunction issued by the Cincinnati court restraining EMF from selling its own cars would remain in force. Judge Swan did accept the EMF formal response to the original bill of complaint which the company submitted at this time.

EMF Replies to the Studebaker Suit

In its response, EMF claimed that it had canceled its contract with Studebaker only after the latter had repudiated it by refusing to accept the agreed-upon number of cars. It maintained that the majority of the EMF directors had supported this action after a meeting in which a vote had been taken. Moreover, EMF stood to lose the most, not Studebaker, if the Studebaker suit were upheld. EMF had accumulated considerable expense from storing hundreds of unsold cars and paying several millions of dollars over the past three months on materials and production improvements.

On the other hand, EMF argued, the Studebaker Automobile Company had little to lose if the sales contract were voided. It did not manufacture any automobiles. It was a separate organization set up by the Studebaker Brothers Manufacturing Company simply to sell automobiles, and it was capitalized for no more than $100,000. Therefore, the only losses that the Studebaker Automobile Company could incur by the cancellation of the sales contract would be those of unrealized automobile sales. However, if the courts enforced the contract, and the Studebaker Automobile Company continued to default on accepting the agreed-upon number of vehicles for sale, EMF would be ruined, and its factories eventually would be forced to close.[7-2]

The action then switched to Kalamazoo. At the hearing on Wednesday, December 22, Studebaker forces had high hopes that this shift in venue would result in the court changing the temporary injunction against EMF into a permanent one.

They were in for a shocking surprise.

Studebaker Is Frustrated in Court a Second Time

On Friday December 24, after hearing testimony from both EMF and Studebaker for two days, Judge Severens ruled that there were insufficient grounds for the suit to be continued and that the injunction to restrain EMF from selling its vehicles was overruled.

Judge Severens' reasons for his decision were both compelling and revealing. He said that the Studebaker suit, as filed, was inappropriate. The complaint was not one between two companies but between the stockholders of the same company; that is, between a minority of the EMF stockholders (Frederick Fish, Clement Studebaker, Jr., and Hayden Eames) and the remainder of the EMF stockholders. The court could interfere only if it could be proved that the EMF board had "willfully deserted its proper functions and perverted its powers by the promotion of schemes mischievous to the corporation. In the present case, no such condition is shown."[7-3]

Severens added that if the Studebaker Automobile Company felt it had been damaged by the actions of EMF, the proper course of action would have been to file suit against the latter company, as it had in Detroit. Severens sent word to Judge Swan that he would be willing to try the Detroit suit for him, but Swan refused the gesture.

Studebaker Requests Dismissal of First Suit

A thoroughly shaken Studebaker Automobile Company returned to the U.S. District Court in Detroit. A new approach was urgently needed. When the hearing opened in Detroit on Monday, December 27, counsel for Studebaker immediately requested that Judge Swan dismiss its bill of complaint against EMF. The EMF counsel, suspecting this was another legal maneuver, asked for a delay in the decision of the court, then returned at 2:00 P.M., and informed the court that it could not agree to a withdrawal of the suit unless the court constrained Studebaker from instituting similar actions in other courts. Judge Swan took the matter under advisement and set Wednesday, December 29, as the day on which he would render his decision.

That Wednesday proved to be another bleak day for the Studebaker legal team. Judge Swan did indeed dismiss the Studebaker bill of complaint against EMF without prejudice. He also added several caveats: that any new suit filed by Studebaker "touching on the subject matter of this suit, or any part thereof" had to be made within the next 10 days and could not be filed in any other jurisdiction other than the Circuit Court in Detroit.[7-4]

Studebaker Files Third and Fourth Suits

Judge Swan's order notwithstanding, on December 30, the Studebaker attorneys marched across the state to the Circuit Court in Kalamazoo and again filed another suit, requesting Judge Severens to grant a new injunction halting EMF from selling its own automobiles. Again, Severens denied the suit.

No one could ever accuse the Studebaker attorneys for a lack of fortitude. From Kalamazoo they immediately set off to Cincinnati, appearing before Judge John W. Warrington of the District Court there on the next day, with the same old arguments plus a new one—that Hayden Eames, general manager for Studebaker, had no right to sign a contract with EMF to accept the number of cars that the latter would produce each month. Again, it would seem that the courts had lost track of the proceedings. Cincinnati intervened and scheduled a hearing for Thursday, January 6, 1910. It was as if Judge Swan's order to restrict all suits to the Detroit court had never existed.

EMF Board Makes Cancellation of the Studebaker Sales Contract Official

On Friday of this turbulent week of December 27, the EMF board of directors broke the action by holding a special meeting at the request of its president, Walter Flanders. He reviewed all events that had taken place, leading up to his letter of December 9 to Studebaker which cancelled the contract between the two companies. He then offered a resolution, asking the board to ratify, approve, and confirm the contents of the letter, thereby being on the record as having severed all business relations with Studebaker. Frederick Fish protested the resolution, but the four non-Studebaker directors, being in the majority, carried the day. Try as he might, Fish could not convince the non-Studebaker forces to forge a compromise.[7-5]

EMF Now Sues Fish, Eames, and Studebaker, Jr.

With this bit of by-play behind the company, the EMF attorneys immediately drove into downtown Detroit and filed suit in the Wayne County Circuit Court, asking the court to remove Messrs. Fish, Eames, and Studebaker, Jr. from their positions as directors of EMF. Why? Because the three were conspiring to bring about the downfall of EMF so that "it might be purchased at advantageous terms by the Studebaker Automobile Company."[7-6] Now the entire issue was brought into the open.

Interestingly, also on that same day, Studebaker informed its dealers, whom it earlier had forbade from doing business with EMF, that the company had changed its policy, and they could order their cars directly from EMF if they felt they might suffer financial loss by not doing so. As a result of this decision, the Studebaker Automobile Company was placed in the embarrassing position of seeing many of its dealers journeying to Detroit to compete with other independent agencies over the right to represent EMF in the same city. Generally, the EMF sales department favored the Studebaker agents, knowing how badly they had been hurt by the sudden withdrawal of EMF vehicles from their sales rooms. Indeed, during the last week of December, EMF shipped approximately 600 cars to its newly signed-on dealers. In fact, the company had received orders for 10,000 cars since publishing the December 9 letter rescinding the sales contract with Studebaker.[7-7]

With the coming of the New Year, one would think that both companies by now would have exhausted their legal struggles. Such was not the case.

On January 3, 1910, the Studebaker attorneys were in Judge Swan's courtroom for the fourth time to file another suit against EMF—this one for $2,000 in damages. Again, they requested Judge Swan to grant an injunction to prevent EMF from selling its automobiles. Although Judge Swan declined to issue such an injunction, he did accept the suit. However, he did not set a future court date for a hearing.

Cincinnati Court Rips into Studebaker

On Monday, January 10, the Cincinnati hearing was held. Studebaker had added a few new wrinkles to this suit, now claiming that it had spent $1 million in advertising the sale of Studebaker-EMF cars and had paid EMF a total of $46.2 million over the past three years for the rights to be the sole purchaser of EMF auto production. The language in the new suit did not fool Judge Warrington. He excoriated the Studebaker attorneys for taking up his time. He said that the current bill of complaint offered nothing new. Moreover, it was improper for a second judge to grant what a first judge of similar standing (i.e., Judge Swan) had refused. Judge Warrington apologized for allowing the hearing to take place and refused to hear the case.[7-8]

At this point, Studebaker was down to a single suit—the one it had filed in Detroit on January 3. Things were not looking good, especially when the EMF attorneys brought in supplemental affidavits on the following day in response to the latest suit in Detroit.

EMF Condemns Studebaker Sales Practices

EMF used the affidavits to attack Studebaker for providing its retailers only a 10 percent commission when they sold EMF vehicles. In one affidavit after another from such industry luminaries as Will Durant of General Motors, Robert Hupp of the Hupp Motor Car Company, Wilfred Leland of Cadillac, Ben Briscoe of Maxwell, Frank Briscoe of the Brush Runabout, and others came testimony that an auto manufacturer could not exist if it allowed its selling agents only a 10 percent profit. In general, they said a profit of approximately 20 percent was the norm. Furthermore, EMF had been paying its own agents on the East Coast an average of 23.5 percent commission until September 1, 1909, when the responsibility for Eastern sales was taken over by Studebaker.

In another affidavit, Walter Flanders claimed that Studebaker injured the reputation of EMF automobiles by placing the name "Studebaker" in large letters above the name "E-M-F" on the nameplate when the cars were sold.

Finally, the EMF answer added that the company had failed to realize the pitfalls contained in its contract with Studebaker at the time the contract was signed, but had terminated the contract when it "saw ruin staring it in the face when the full consequences of the agreement became evident."[7-9] Another addition was a bill for $10,000 for the storage of vehicles that Studebaker had refused to accept.

On Wednesday, February 10, 1909, Judge Swan dismissed the Studebaker petition and refused to grant an injunction against EMF.[7-10]

Undeclared Truce Brings Quiet Time

After five weeks of incessant legal turmoil, a period of peace ensued. No new suits were filed. No new hearings were held. It was as if the two companies had arrived at a stand-off. However, beneath this veneer of tranquility, the plot was thickening as Walter Flanders continued to search for any means he could muster to better the position of EMF within the industry.

Early in February, Flanders journeyed to New York to discuss the possibility of EMF joining the United States Motor Company that Ben Briscoe was in the process of pulling together. This auto conglomerate, expected to be a rival to General Motors, was announced to the public on January 26, 1910. Briscoe already had merged the Maxwell-Briscoe Motor Company and the Columbia Motor Car Company, and was in the process of adding the Dayton Motor Car Company, the Brush Runabout Company, and the Alden-Sampson Manufacturing Company. It was quite a team and might have been well served by the addition of EMF, even more so with the addition of a person with the drive and expertise of Walter Flanders (as events later proved).[7-11] Nothing came of a Flanders/Briscoe venture, however.

Meanwhile, through January and February 1910, Flanders continued his efforts to maximize the output of both the EMF 30 and the Flanders 20. Production was boosted to 100 chassis per day when he and Max Wollering installed what might have been the first mechanized assembly in the auto industry. The line had approximately 30 workstations and was controlled by a clerk who recorded the order and completion of work at each station before the line was moved forward.[7-12]

For all the legal barbs that Studebaker kept hurling at EMF, one would think that the principals involved would have had difficulty remaining in the same room together. Actually, they remained friends. Most of the action seemed to take place between Flanders and the Studebaker legal team, whose lawyers continued to report to Frederick Fish as chairman of the Studebaker executive committee. They had little encouragement to offer Fish as time passed and were of the opinion that Flanders had made up his mind not to settle with Studebaker—in fact, that Flanders would prefer to continue to fight them. In turn, Fish kept reminding his lawyers to keep applying the pressure.

Meanwhile, EMF continued selling its own cars, much to the chagrin of Studebaker, many of whose own agents were doing the selling but for their own profit rather than for the profit of Studebaker. The longer that matters went unresolved, the more income Studebaker failed to realize. A rough guess would be that Studebaker was losing more than $200,000 per week through unrealized sales, assuming that EMF was selling every car that it built each day.

Fish Brings in a Hired Gun: J.P. Morgan

Seeing no way to achieve his purpose through legal means, Frederick Fish decided that the most promising course of action for Studebaker would be to make an offer to buy out EMF—an offer that would be so outrageous it could not be refused. However, Fish would do so through a third party: the House of Morgan in New York. It was a good move. J.P. Morgan & Company had been involved in Studebaker affairs since 1894 and owned a 17-percent interest in its stock. Also, Morgan, through its dealings with Ben Briscoe and others, was fast becoming familiar with the business needs of newly established auto companies.

Midway through February, the Morgan legal minds and the EMF board began a series of intensive negotiations that ultimately bore fruit. By the end of February, an informal agreement had been reached that all outstanding stock in EMF, except those shares owned by Studebaker, would be purchased by J.P. Morgan & Company.

On Saturday, March 5, having been apprised of the progress made by Morgan in purchasing EMF, the directors of the Studebaker Brothers Manufacturing Company met and voted to terminate the bill of complaint that the company had pending against EMF. The board also released EMF from any damages that might have resulted from the legal action.[7-13]

On the following Tuesday, March 8, the EMF board of directors met secretly at 3:00 P.M. in the offices of J.P. Morgan & Company on Wall Street, at which time the terms of the sale of EMF to the Morgan interests (representing Studebaker) were announced. Board members present were William Barbour, Frederick Fish, Clement Studebaker, Jr., and Walter Flanders. A reorganization of the board was in order. The resignations of James Book as vice president and director of EMF, Hayden Eames as director, Charles Palms as treasurer and director, and William Barbour as director were tendered and accepted. In their places as directors were elected Frederick Stevens, F. Gordon Brown, Frederick Delafield, and Francis H. McKnight. Each of the new directors was a Morgan man. Brown then was made vice president, and Robert M. Brownson, a Studebaker man, was named treasurer. A new contract was awarded to Flanders, hiring him as president and general manager for a three-year term retroactive to January 1, 1910.[7-14] EMF now belonged to Morgan and Studebaker, although the Studebaker ties would not immediately be made public.

J.P. Morgan & Company Announces Purchase of EMF

Now it was the public's turn to learn of the change in the EMF guard. On March 9, 1910, Detroiters awoke to huge headlines announcing that the Everitt-Metzger-Flanders Company, the largest employer in the city, had been sold to J.P. Morgan & Company. Each of the three leading Detroit newspapers offered a different view of the sale.

The Detroit News, for example, trumpeted the money that the EMF principals realized from their original investment. "BIG FORTUNES MADE IN E-M-F STOCK BY DETROIT OWNERS," its headline read. Much attention was devoted to the fact that the wealthy Dr. James Book, the leading stockholder of EMF who had held title to almost 25 percent of its stock, was heralded to have cleared approximately $1.25 million in the transaction from his original, 18-month investment of approximately $120,000.

"E-M-F COMPANY BOUGHT BY MORGAN INTERESTS" was the headline of *The Detroit Free Press.* Its emphasis was placed on the fact that Walter Flanders would remain as EMF general manager and one of its directors, and that Flanders would retain a substantial block of stock in the "new" company. In effect, the article stressed that EMF would continue to conduct its business as in the past, both in manufacturing and at the sales end. The only exception in the company routine would be the new faces that appeared on the EMF board. Not specifically mentioned was that Walter Flanders' original contract had been torn up and replaced by a new one for three years retroactive to January 1. It would pay him $30,000 per year plus one percent of the net earnings of EMF if they exceeded $2 million.[7-15]

The Detroit Journal came closer to the truth of the transaction than its rival newspapers with its headline, "PURCHASE OF E-M-F AUTOMOBILE CO. FINAL COUP BY STUDEBAKERS." Unlike *The Detroit News* and *The Detroit Free Press*, *The Detroit Journal* printed a statement issued by J.P. Morgan & Company, which read:

> Pursuant to an arrangement with the stockholders of the Studebaker Bros. Manufacturing Co., J.P. Morgan & Co. have purchased substantially all of the stock of the Everitt, Metzger, Flanders Co. of Detroit not already held by the stockholders of the Studebaker Bros. Manufacturing Company. In connection with the purchase, a contract has been entered into with Walter E. Flanders to continue as president and general manager of the Everitt-Metzger-Flanders Co. for a period of three years. As part of the transaction, the litigation pending at Detroit is terminated.

The Detroit Journal also wrote of the kind remarks about Walter Flanders made by the Morgan interests.

> All through the litigation between the Studebakers and the E-M-F Company, the name of Walter Flanders was never mentioned in a disparaging way. Mr. Flanders is a wonderful producer, and his services to the Studebaker Company would be too valuable to turn away.

Although various purchase prices for EMF were quoted among the media, it became generally accepted that Morgan paid approximately $6 million, or $60 per share, for the EMF stock. Inasmuch as the going price for EMF stock was $35 per share, this indeed represented a daunting offer that the current EMF directors would have been hard pressed to ignore, especially when their original investment came to only $10 per share. One estimate reported that EMF stockholders received $32 for every $1 they invested when the company was formed.[7-16]

All three newspapers speculated whether the EMF purchase was the first step in a plan by which J.P. Morgan & Company would form another giant trust to rival the likes of U.S. Steel, which that firm had brokered several years previously. In 1908, Morgan had been party to the negotiations that ended in the formation of General Motors, and only the previous month another auto conglomerate had been announced: Ben Briscoe's United States Motor Company. Would EMF become the foundation stone for a third? *The Detroit News* reported that Wall Street thought so, especially after its members discovered that the House of Morgan was interested in providing financial backing to the Wright brothers as well as buying EMF. Soon, Wall Street predicted, it could very well be possible that "One must travel by air, earth, or water over a Morgan line; speak over a Morgan wire; draw one's checks on a Morgan bank, and when one dies, the widow will get the insurance money from a Morgan company."[7-17] This sequence of events, of course, never took place, although later, in 1920, the House of Morgan would come to the rescue of General Motors after Durant brought it to the point of bankruptcy.

Much was made of the financial gain realized by the original EMF stockholders, especially that of James Book, and that Studebaker now had become the owner of a very successful automobile company (one that would make it a significant figure within the auto industry over the next four decades). However, the man who personally realized the most profit from the sale of EMF to the Morgan and Studebaker interests was Walter Flanders.

Walter Flanders: The Horatio Alger Story of 1910

Walter Flanders now was a very rich man—a millionaire. It had taken only five years for him to reach this lofty level. Five years ago, Flanders had been only one of many machinery salesmen plying his trade. One year later, he had impressed Henry Ford sufficiently to have Henry put him in charge of Ford auto production. Two years later, Flanders left Ford and helped found a company called EMF, to which he contributed nothing more than his name, expertise, and driving ambition. Flanders neither was one of the original directors of EMF nor initially owned any of its stock, but he was paid a salary of $6,000 per year as general manager. Under his direction, EMF had become the largest employer in Detroit, producing more automobiles than all but two or three auto manufacturers in the entire United States. Now Flanders was independently wealthy from selling stock that had been given to him in reward for his services. Not only was he independently wealthy, but he had an iron-clad contract to continue as president and general manager of EMF for the next three years at the princely salary of $30,000 annually. He also was in charge of its advertising and a sales force numbering 1,000 dealers.

Moreover, Flanders continued to have the solid support of his key management team—members such as Max Wollering and LeRoy Pelletier who periodically had been awarded shares of stock by the EMF board over the previous year for their high performance. This stock was valued at $400,000 to $600,000 in total when Morgan purchased the company. For example, LeRoy Pelletier's share was said to have come to $175,000.[7-18] In fact, during negotiations with Morgan, Flanders intentionally arranged with the EMF board so that none of the directors would attempt to take advantage of inside knowledge and buy up small blocks of EMF stock. In this way, both the EMF directors and its management would benefit equally from the stock sale.

After the sale and after it was announced that EMF would continue to do business as usual despite new ownership, Flanders proudly lauded his organization. He exclaimed to reporters,[7-19]

> I have today what I call "the greatest commercial fighting machine" in this industry.

Although EMF may have become a satellite of the Studebaker firm, one would have been hard pressed to realize that such a relationship existed when listening to Flanders talk about his plans for the future of EMF. In advertising and marketing the EMF product, Studebaker was scarcely mentioned.

Studebaker Takes Over EMF, Flanders Remains in Control

To be truthful, Frederick Fish had placed the Studebaker Brothers Manufacturing Company in a precarious position. He had been forced to borrow $3.5 million to finance the EMF purchase using most of the Studebaker family collateral to do so. Moreover, the House of Morgan, which had first call on the $3.5 million in notes due in 1911, forced Studebaker to place two of its men, Stevens and Delafield, on the Studebaker board of directors. This was not surprising because Morgan owned one-sixth of the outstanding stock in Studebaker initially and now held notes on millions of dollars of Studebaker property, but had not as yet had a seat on the board to represent its views.

Stevens and Delafield were elected to the Studebaker board at its meeting of March 23, 1910. Also added was Walter Flanders, head of its EMF subsidiary. To make room for them, Messrs. Carlisle, Innis, and Studebaker, Jr.—all Studebaker family members—had to resign. It was the end of an era as Studebaker family members saw their power over company affairs diminish considerably.[7-20]

As a separate satellite of Studebaker, the EMF board continued to exist as an independent entity and deliberate on the business of that company. However, all of its decisions then were passed on to the Studebaker board for approval.

Flanders' Control Appears Stronger Than Ever

On April 5, Flanders presented his ambitious production schedule to the EMF board. It called for the manufacture of 7,400 to 9,200 EMF 30's between April and the end of December 1910, as well as 11,600 to 13,400 Flanders 20 models.

He also surprised everyone by announcing that there would be an EMF 35 ("35" denoting horsepower) that would go into production on September 1, and he expected to build 1,400 of them by the end of the year. The board approved the new schedule.

Flanders now projected his estimate of EMF production for 1911, which came to a grand total of 33,850 to 38,550 vehicles. (The audacity of the plan can be measured by the fact that it would have exceeded the number of cars Ford eventually built in 1910.) Now that Flanders had the attention of the board, he also added that he intended to reduce the price of the EMF 30 from $1,295 to $1,150 by September 1 through savings in its cost of manufacture. As a final shock, Flanders told the directors that after the first of the year, EMF would begin to develop a six-cylinder car for the 1912 model year![7-21]

One bit of business remained: to rehire James G. Heaslet to a three-year contract as chief engineer of EMF, at a starting salary of $8,000. This would be increased to $17,000 in 1911 and to $20,000 in 1912.

The Studebaker directors met on April 9, 1910, and approved the actions of the EMF board to formally place authority for all advertising and sales of EMF cars in the hands of Walter Flanders. The 19 branches of the Studebaker Brothers Manufacturing Company now were under contract to report to Flanders directly in all matters relating to the sale of automobiles, including the Studebaker-Garford (which sold for three times the price of an EMF 30) and its electrics.

By July 1910, Studebaker had severed its ties with Garford.[7-22] During their seven-year alliance, fewer cars had been made and sold in total than normally would have been built in two months by EMF.

According to new terms worked out by Flanders with Studebaker approval, EMF/Studebaker dealers were required to keep an adequate supply of EMF 30 and Flanders 20 models on hand and could sell them only at the fixed price of $1,250 for the EMF 30 and $750 for the Flanders 20. The dealers in turn purchased the cars from EMF at discounts of fifteen percent and eight percent, respectively. The dealers also were to purchase and maintain a sufficient supply of parts and accessories to deal efficiently with subsidiary dealers and EMF customers. Moreover, when expanding into new territories, sales agents were to contract directly with EMF rather than through Studebaker.

Interestingly, Studebaker also consented to a motion earlier passed by EMF that the name "Studebaker" would not appear on the vehicles. The EMF sales policy of designating EMF automobiles as the E-M-F "30" and the Flanders "20" was continued, and the name "E-M-F Company" was to appear prominently in any advertisements.[7-23] Thus, although Studebaker now owned EMF, it allowed the latter company to proceed as if it were an independent business unit. A glance at the EMF ads published during 1910 verifies this. No mention is made of any affiliation between EMF and the Studebaker Brothers Manufacturing Company.

Nevertheless, the purchase of EMF was paying off handsomely for Studebaker. A report circulated in *The Detroit Journal* on July 20, 1910 estimated the production of EMF 30 cars for 1910 to be 14,300 units and of Flanders 20 models to be 23,000. Together, they represented approximately 27 percent of all new cars built that year by the 23 auto firms located in Detroit. For a company that was only two and one-half years old, this was quite a feat.

On a less positive note, E. LeRoy Pelletier, the imaginative advertising manager of EMF who had kept that title at a salary of $20,000 per year, submitted his resignation on May 27 on the eve of taking a trip to Europe. Rumor had it that J.P. Morgan had become angered because he felt that Pelletier had been tossing his name around too freely in connection with EMF and Studebaker affairs and wanted Pelletier removed from the company.[7-24] This hardly seems likely. If it were true, Pelletier would never have been able to sign on as advertising manager for Studebaker after his return from Europe, which he did.

By August 1910, the sales of EMF cars and those of the Ford Motor Company and other auto companies were proceeding at a merry clip. However, horror stories began to circulate from some bankers that folks were mortgaging their homes and farms to obtain the money to buy a motorcar. *The Detroit Journal* asked Walter Flanders whether this was true. Flanders understood the question to be one prompted by people buying cars on the installment plan. He called the bankers' rumors nothing more than gossip except for those instances where customers were dealing with "mail-order cars and mushroom firms whose factories are merely miniature assembly plants." Flanders added that never in his own experience had he known a person who had mortgaged his house to buy an automobile, and that at least nine of every ten cars were sold on a cash basis. Flanders did surprise *The Detroit Journal* readers by saying,[7-25]

> The prosperous Western farmer is now by all odds the largest buyer in the market. Nearly 90 percent of our 1910 output has gone into the hands of owners living on farms or in small villages, where streetcar lines do not exist.

The year 1910 was, in a sense, a golden year for Walter Flanders. He was rich, and he had unlimited authority over one of the largest and most productive auto companies within the industry, now a subsidiary of powerful Studebaker. The EMF factories representing a $7 million investment—all paid for—were building 80 EMF 30 cars and 125 Flanders 20 cars every working day (according to the EMF ads). Flanders' young company had appeared from nowhere to become the primary threat to Buick, Willys, and Ford for supremacy in the automotive industry. His factories employed 12,500 men, whom he considered the "pick of the trade." If any flaw existed within his forward planning, it was Flanders' conclusion that he had made such a perfect car in the EMF 30 that nothing remained upon which to improve.

Problems with the Flanders 20—and Solutions

Flanders may have been right about the EMF 30, but the Flanders 20 was another matter. The Heaslet design had incorporated a two-speed transmission, the strain from which often led to a snapped axle. The company was plagued with complaints from the field. A shift to a three-speed transmission on 1911 models cured the problem.[7-26] Other than that, the Flanders 20 had much to offer. It featured several items from the more expensive EMF 30, such as a Splitdorf magneto, two sets of internal expanding brakes, and a full elliptic rear spring with scroll ends and extra-heavy spring clips. Also copied from EMF 30 practice were a drop-forged camshaft with integral cams, a float feed carburetor, vacuum feed lubrication, a uniquely designed rear axle in which the transmission was incorporated, and a "no backpressure" muffler.

Driving around the town in an EMF 30, circa 1910. (Courtesy of the Detroit Public Library, National Automotive History Collection)

Customers could choose from four different Flanders 20 models. This was in keeping with Flanders' desire to sell as many of the low-priced models as possible. There was a Runabout for doctors and salesmen that sold for $700. Also at $700 was the Racy Roadster model with long fenders and low seats. The Suburban model had a detachable rear seat that could be converted into a truck-type bed for carrying light loads, and it was priced at $725. Most expensive was the Coupe at $975, but it had an enclosed body. Selling at such low prices, the Flanders 20 was poised to make a serious run at the Ford Model T.

EMF conducted a vigorous publicity campaign to convince the public of the durability and reliability of the Flanders 20. For example, on June 6, 1910, a stock EMF 20 was dispatched from Quebec, Canada, on a 4,127-mile run to Mexico City, where it arrived on August 3—none the worse for wear. In fact, at times the car averaged 30 miles of travel on a single gallon of gasoline.[7-27]

Early in December, a four-passenger 1911 Suburban model Flanders 20 was taken off the line and transported to the Pacific Coast, where it was driven nonstop on the streets of Los Angeles from December 2 through December 30, having set a record for continuous operation of 10,872 miles. An observer from the Automobile Association of Southern California was onboard at all times to ensure the veracity of the claim.[7-28]

The following year, company advertising hailed another Flanders 20 feat—a 1,281-mile jaunt from Seattle to Hazleton, British Columbia, over roads that allegedly had never seen a motorcar. P.E. Sands, the Seattle EMF/Studebaker dealer, dared to make the trip with two others. The price people pay for a bit of publicity! Following is some of the advertising copy that related to the trip:

> Through brush, streams and mud, over boulders and every conceivable impediment, the Flanders "20" kept steadily going. Sometimes it seemed as if the car could never succeed; once Sands was almost killed as the soggy trail gave way and tipped the car over. But the Flanders kept at it, pulled over 500 miles in low gear and came into Hazleton one bright night with her motor humming as sweetly as at first.

Both the Flanders 20 and the EMF 30 received good reviews during their appearance in the important January 1911 New York Auto Show. The EMF 30 was pointed out to have two new models to add to the touring car that was the mainstay of EMF sales from 1908 through 1910. One was a roadster with two rear bucket seats mounted over a large gasoline tank. The other was a demi-tonneau (removable rear seat design) in which the tank was moved beneath the front seat.

Studebaker Finances Are Backed Against the Wall

The problem faced by Studebaker because of its costly purchase of EMF finally came to roost. The roster of EMF stockholders clearly signaled where that problem lay. In the annual report filed by EMF with the state of Michigan on November 28, 1910, the list indicated that the Studebaker interests (Fish, Studebaker, Jr., and Flanders) owned outright only 191 shares of EMF stock. The four Morgan representatives had title to the remaining 99,809 shares.

Presumably, these shares were being held in trust (or as collateral) for Morgan because the latter had formed the consortium by which Studebaker was able to borrow $3.5 million in notes to buy out EMF. Morgan held the first option on these notes, which were coming due in March. The notes were in the form of bonds placed on all Studebaker properties.

Out of the blue, J.P. Morgan & Company decided in late 1910 that it would not exercise its option to purchase the bonds, which meant Studebaker would have to pay up when they came due. The sudden financial collapse of General Motors in September 1910 and its subsequent rescue by a consortium of Eastern bankers had caused Morgan to rethink its own views about the stability of the fledging auto industry. It would be another decade before J.P. Morgan & Company seriously became involved in automobile affairs with—whom else?—General Motors.

Fish knew that Studebaker did not have the funds to pay the due notes and had counted heavily on a sympathetic hearing from its old friend, the House of Morgan, to purchase and hold them. With Morgan now out of the picture, Studebaker was in dire jeopardy. It conceivably would have to sell off almost all family assets to raise the necessary cash; otherwise, it would be forced to put EMF back on the block without having had the opportunity to capitalize on its earning power.

Fish Forced to Negotiate with Eastern Bankers Again

Before moving to South Bend after marrying Grace Studebaker, Frederick Fish had been a prominent corporate attorney in Newark, New Jersey, with close ties to Wall Street. It was there that he headed for financial relief. Fish found his answer in Henry Goldman, founder of Goldman, Sachs & Co., for whom Fish had once worked. Goldman decided to rescue Fish. He put together a new consortium that purchased the $3.5 million in outstanding notes. Friend or not, Goldman extracted a stiff price for the loan—the reorganization of Studebaker from a wagon maker to an auto company.[7-29]

Thus, on Valentine's Day, February 14, 1911, the Studebaker Brothers Manufacturing Company became the Studebaker Corporation capitalized at $15 million in preferred stock and $30 million in common stock. Of the

$15 million in first preferred stock, Kleinwort Sons & Co. of London purchased $13.5 million in conjunction with Goldman, Sachs & Co. and the Lehman Brothers of New York. The influx of new money allowed Studebaker to pay the notes and remain solvent.

However, demands made by the owners of the $13.5 million in preferred stock were severe, to say the least. In simple terms, their stock would receive a seven-percent dividend yearly, and they would have first lien on Studebaker assets should the company fail. No dividends would be paid on the $30,000 of common stock until the company had built up a reserve of $1 million, after which a six-percent dividend could be declared (but no more) until the reserve had been built up to $2.5 million. Furthermore, none of the Studebaker property or assets could be mortgaged without the consent of 75 percent of the stockholders.[7-30]

Henry Goldman and Paul Sachs were elected to the Studebaker board in place of Frederick Stevens of J.P. Morgan & Company and supervised their interests quite closely, requesting and receiving weekly reports from South Bend on all aspects of its business—even to the salaries of management.

EMF Following the Studebaker Reorganization

For Flanders, the big change coming from the reorganization of Studebaker was that the EMF identity gradually became submerged within the Studebaker culture. Its independence was taken away. He did receive a new five-year contract to act as general manager of the "Automobile Department" at a salary of $30,000 per year and one percent of net earnings that exceeded $2 million.[7-31] However, whereas EMF cars previously had been advertised as being manufactured by the E-M-F Company with scarce or no mention made of its ties with Studebaker, the EMF cars now were advertised as products of the Studebaker Corporation below which the words "E-M-F Factories" usually appeared in very small print.

Studebaker may have backed off from promoting the EMF title, but it did attempt to capitalize wherever possible on the Flanders name. For example, a lengthy seven-page ad appeared in *Munsey's Magazine* in 1911, with no other purpose than to extol the virtues of Walter Flanders. Beneath its headline, it read[7-32]:

The True Wonder-Story of the Poor Vermont Machinist who turned $195,000 into $6,000,000 in twenty months, made his associates rich, placed his automobiles on every highway in the union, and shared his results with his men.

New Flanders 20 Introduced

In March 1911, shortly after the new Studebaker Corporation was formed, EMF announced that a new five-passenger Flanders 20 selling for $800 would be in production by April. It would have a longer wheelbase (102 inches vs. 100 inches), a detachable exhaust manifold, a stronger rear axle, and brakes with double the face width of the previous car.[7-33] What would distinguish this model from most other U.S. cars of that year was that it would be equipped with the "fore-door look;" that is, it would have both front and rear doors. (Four-door cars would quickly increase in popularity and be offered by virtually all other manufacturers by 1913.)

As advertising manager, Pelletier made certain that the Flanders 20 received copious exposure in the media during 1911, not only through advertisements but also indirectly through news stories of its exploits. For example, a Flanders 20 won the rigorous 1,390-mile run from Twin Cities to Helena, clipped 47 seconds off the record for climbing Dead Horse Hill in Massachusetts, finished first in a reliability run between St. Louis and Kansas City, and took the Newport Hill Climb up the worst hill in Indiana. It also acted as the Pathfinder car for the 1911 New York to Jacksonville Glidden Tour.

Recipients of trade journals such as *The Motor World* in June 1911 were treated to one of the longest automobile advertisements to appear anywhere—48 pages of photos and copy from Studebaker Corporation under the banner, "Which is the Largest Automobile Factory in the World?" In the advertisements, the company repeatedly made the case that the

> Studebaker Corporation-E-M-F plants are the largest in the world...They are the largest in number of cars produced—over 30,000 per annum. They are the largest in volume of business done—over twenty-five millions ($25,000,000) of dollars per annum. They are the largest in the amount of money invested in factories and equipment—over $8,000,000.

A 1912 Flanders 20 runabout, as it appears today in the Studebaker Museum.

The advertisements dwelled on the fact that every part for every Studebaker was made, tested, finished, and assembled into an automobile, all within the same factory, then sold through its own agencies. They made a strong case for a massed-produced car whose parts were interchangeable, therefore was stronger and better built than handcrafted cars, and, by virtue of the larger numbers built, could be manufactured more economically and thereby sold at a lower price.[7-34]

Features of the New EMF 30 for 1912

Early in July 1911, EMF began building a modified EMF 30 for the 1912 model year. In an ad that appeared in *The Motor World* on August 3, 1911, the company claimed that it already had delivered 5,000 of the EMF 30 even though the car had not yet been formally introduced. Physically, the new EMF 30 came on a 112-inch wheelbase, eight inches longer than that of models in the previous year, and it featured drop frame construction. More importantly, the EMF 30 also had adopted the "fore door" or full-door look.

This 1912 EMF 30 "fore-door" touring car, shown here at a recent auto show, was originally priced at $1,100.

The magazine, *Cycle and Automobile Trade Journal*, described the new EMF 30 in rather eloquent terms[7-35]:

> The E.M.F. "30" vestibuled [four-door] body is designed to carry out the most beautiful and effective ideas of fore door construction. In the dash has been built adjustable ventilators, giving free circulation of air in summer, or closed for warmth in the winter. The dash is of the semi-torpedo design and harmonizes with the wide sweeping lines of the fenders and the graceful bodylines.

Crisis or Comedy?

The preliminary catalog for the 1912 Studebaker EMF 30 shows how EMF cars were being advertised as products of the Studebaker Corporation. (Courtesy of the Detroit Public Library, National Automotive History Collection)

The E-M-F Company

A 1912 EMF 30.

What began as a year of turmoil for Studebaker ended as a total success for EMF as the latter catapulted into second place in the industry, with more than 25,000 units having been driven out of its factory. This truly was a startling achievement for a company that had begun on a shoestring but, under the forceful personality and skill of a man such as Walter Flanders, had stunned the rest of Detroit with its rapid rise to success.

In one of the ironies of the trade, Flanders purchased in July the former Ford plant on Piquette Avenue, a block east of the EMF headquarters where he had begun his meteoric career, as he expanded his operations. The plant had become empty not long after Henry Ford had moved Ford assembly operations to Highland Park. Flanders expected that the Piquette space would help him increase EMF output by another 20 percent.[7-36]

However, by the fall of 1911 and despite the enormous success being enjoyed by EMF automobiles, rumors began circulating among insiders that Walter

Flanders was looking for a way out of his contract with the Studebaker Corporation. Such talk no doubt stemmed from Flanders taking the unusual step of organizing a new company under his own name while continuing to act as a vice president of Studebaker and general manager of its automotive division. What is equally unusual is that Studebaker allowed Flanders to do this. For the account of how all this took place, we must turn back the calendar.

Flanders' Folly: The Great One Divides Himself

During May 1910, with no apparent worlds left to conquer, Walter Flanders convinced James Book and William Barbour, his previous EMF money sources, to join him and A.O. Smith, the Milwaukee frame-maker, in a new enterprise. First, they purchased the Grant Automatic Machine Co. of Cleveland, then moved its equipment into a building bought from the Chelsea Stove Works. They called their new firm the Grant & Wood Manufacturing Company and capitalized it for $1 million. John Grant, the previous owner, was retained as a consulting engineer. Although not officers in the company, Book, Barbour, and Flanders were on its board of directors.[7-37]

Five months later, another new company was formed with which Flanders' name was closely intertwined—the Pontiac Motorcycle Company. It was capitalized for only $600,000 and would not have made much news, except that its chief stockholders were identified as James Book, Charles Palms, and Walter Flanders. What was even more surprising was that Robert M. Brownson had resigned as secretary and treasurer of EMF to become the president of the new company, accompanied by Max Wollering, the EMF production manager, who took the same job with Pontiac. These were two of Flanders' most trusted confidants, and their departure from EMF for a smaller, relatively insignificant company naturally gave rise to all sorts of rumors, especially that Flanders himself was about to leave Studebaker.

Flanders discounted the rumors vigorously. He said,[7-38]

> I do not consider that my task in the automobile industry has been fulfilled. In fact, I feel that we have just begun. The E-M-F Company is today the third largest in the licensed association, and it is my ambition to make it the first...There is

no truth in the reports to that effect which have been circulated by competitors of the E-M-F Co.

Nevertheless, over the next few months, the same names began cropping up again in association with the formation of several other Pontiac companies—the Pontiac Drop Forge Company, the Pontiac Foundry Company, and the Vulcan Gear Works. It also was no secret that Walter Flanders was the "father" of all five companies.

Enter the Flanders Manufacturing Company

On Tuesday, January 10, 1911—two months before Studebaker was forced to reorganize—the final shoe in this bit of drama fell. On that date, a new entity was announced to the Detroit scene—the Flanders Manufacturing Company, a merger of five smaller manufacturing companies into one large one beneath the Flanders banner. The five companies? The Grant & Wood Manufacturing Company of Chelsea, the Pontiac Motorcycle Company, the Pontiac Drop Forge Company, the Pontiac Foundry Company, and the Vulcan Gear Works, of course.

The stated purpose of the new Flanders Manufacturing Company was to manufacture automobile parts as well as a new type of motorcycle called a "Bi-mobile," which would be accompanied by a light delivery vehicle called the "Tri-mobile." Both the Bi-mobile and Tri-mobile would be built in the Pontiac Motorcycle Company plant; the other Pontiac facilities would specialize in auto parts for which they had a ready market in Studebaker by virtue of Flanders making the decisions for both companies. The Chelsea facility would focus on the manufacture of automatic, multi-spindle screw machines, screw machine products, and ball bearings.[7-39]

What made the venture so intriguing was the size of its capitalization and the principals involved. As filed in its annual report to the state of Michigan at the end of the year, the Flanders Manufacturing Company was capitalized at $2.25 million, of which $1.75 million represented common stock and $0.5 million preferred stock, par value $100 per share. This was more than twice the capitalization of the now highly successful EMF Company!

Crisis or Comedy?

The exterior of the Flanders Manufacturing Company in Pontiac, Michigan, as it appeared in 1911. (Courtesy of the Detroit Public Library, National Automotive History Collection)

Robert M. Brownson was elected president of the Flanders Manufacturing Company, A.O. Smith of the A.O. Smith Steel Works became vice president; James Book, treasurer; and Harry Stanton, secretary. Directors were Walter Flanders, James Book, William Barbour, A.O. Smith, John Shaw, and, of all people, Clement Studebaker, Jr. We do not know what Frederick Fish at Studebaker must have felt about one of the Studebaker family members investing $400,000 in a rogue company in order to become one of its directors. Nor do we know whether he appreciated that the president and general manager of his most successful operation, the EMF Company, was in a sense "moonlighting," although he did not openly oppose Flanders' action. In fact, Flanders ultimately hijacked Fish into making Studebaker part of the Flanders Manufacturing Company action.

Flanders Tricks Studebaker One More Time

How Flanders did so was a marvel of business maneuvering. When EMF and Studebaker were merged into the Studebaker Corporation in February 1911, Henry Goldman and other members of the Studebaker executive committee offered Flanders a new contract as general manager of the automotive operations. In a letter dated April 24, 1911 that he forwarded to the Studebaker board, Flanders agreed to the terms proffered. However, he made one key addition—that Studebaker also exchange the 2,000 shares of preferred stock and 3,000 shares of common stock that he owned in the Flanders Manufacturing Company for 2,500 shares of preferred stock and 2,500 shares of common stock in the Studebaker Corporation. Par value of the Flanders Manufacturing stock was $100 per share. This trade, he argued, would enable Studebaker to benefit from the wide variety of parts that the Flanders Manufacturing Company could make for the automotive division—parts that were difficult to obtain from other parts companies. Furthermore, Flanders Manufacturing Company was planning to build an advanced, economically priced electric car that Studebaker could substitute for the outdated electric vehicle it now had in production.[7-40]

The closing comments in Flanders' letter were:

> I shall not invest in any other manufacturing company during the term of my contract, because I have decided once and for all to serve the Studebaker Corporation to the best of my ability, and contribute what I can to its success.

This may have been his intent at the time, but later events caused Flanders to ignore his own words.

By convincing Studebaker to buy most of his Flanders Manufacturing stock, Flanders was able to rid himself of more than 75 percent of his holdings in the Flanders Manufacturing Company, rewarding him with a half-million-dollar profit over the next year.

Business must have been good for Flanders Manufacturing Company initially, especially because it had a steady client in Studebaker, thanks to Flanders' manipulation of his contract. The EMF factories had a captured parts source and kept up a steady stream of orders for gears, forgings, and castings. The Chelsea plant also took care of approximately half of the screw machine work for EMF.

The Electric Car That Failed

Flanders Manufacturing Company might have fared quite well had it not made what ultimately became its most fatal move late in July 1911 when it introduced an electric car, a coupe. This strange business venture allegedly was the brainchild of E. LeRoy Pelletier, who himself was "moonlighting" with Flanders Manufacturing Company while continuing to serve as the advertising manager for the Studebaker Corporation.

The Flanders Electric featured worm drive, cradle spring suspension, and extensive use of anti-friction bearings, and it had a wheelbase of 100 inches. It was advertised as a man's car in deference to the marketing of most electric cars that promoted use by "the fair sex." It sold for $1,775, a price advertised as being half that of other closed-body electrics. Much was made of its low wind resistance, light weight, and antique "Colonial" body style with "sashless windows of ground French plate." If the car had been constructed as soundly as the advertising copy suggested and could have been made cheaply enough to give the company a profit at such a low price, it may have been a success. Because neither eventuality occurred, it soon began to flounder.

One also might gather that building the Flanders Electric car was not on the list of duties to which Robert M. Brownson had signed on as president of Flanders Manufacturing Company in January. By the end of the year, Brownson tendered his resignation from the company, having had a serious difference of opinion with Flanders over matters not revealed.

At the next meeting of the Flanders Manufacturing board of directors in December 1911, the board accepted Brownson's resignation and appointed Walter Flanders as president in his place.[7-41] Ordinarily, with Flanders supposedly hard at work running the EMF car division of Studebaker, one would question why he accepted a nomination that might place him in a position where charges of a conflict of interest could be leveled against him.

Surprise—The Big Names Behind Flanders Manufacturing?

Again the rumors began to fly about Flanders leaving Studebaker—with good reason, if one took a close read of the stockholders that made up the Flanders Manufacturing Company. Most were Flanders' old friends. As listed in the annual report to the state of Michigan, giving the status of the company at the end of December 1911, 111 stockholders were on the rolls. Conspicuous among them were the names of the previous directors of EMF; namely, William Barbour, James Book, and Charles Palms. The three men owned 2,351 shares ($235,100) collectively. Max Wollering had invested in 458 shares ($45,800). E. LeRoy Pelletier was in for 832 shares ($83,200).

Other stockholders whose names were well known within the auto industry were Harry Cunningham of the Cunningham Car Company ($30,000), Waldemar Kopmeier of the Kopmeier Motor Car Company ($69,000), Walter O. Briggs of the Briggs Manufacturing Company ($5,000), and A.O. Smith ($35,000). Flanders still owned 2,080 shares ($208,000), all in common stock.

Most surprising to many was the identity of the largest stockholder, none other than the Studebaker Corporation. Studebaker owned 5,000 shares (an investment of $500,000), or 22 percent of those outstanding, having been forced to assume the shares owned by Flanders when he renewed his general manager's contract. With such a sizable amount tied up in the Flanders Manufacturing Company, we might assume that Studebaker had tacitly approved of Flanders' decision to accept the appointment to the presidency of Flanders Manufacturing Company and operate outside EMF while continuing to serve within as the latter's general manager. Either that, or Fish might have had expectations of Flanders working his money-making magic again to preserve the Studebaker investment.

Flanders began as president of Flanders Manufacturing Company with vigor, arranging for a merger between the Universal Motor Truck Company of Detroit and Flanders Manufacturing Company. Universal built a three-ton truck with a wheelbase of 132 inches.[7-42] The company had been founded in 1910.

Rumors continued to grow that Flanders and the Studebaker Corporation were on the verge of parting their ways. Matters were not helped when E. LeRoy

Pelletier announced to the press that the name of the Flanders 20 line was being changed to the Studebaker 20, and that the EMF 30 probably would go the same route and be called the Studebaker 30.[7-43] No doubt Flanders was beginning to feel that his old company was slipping away from him. He may have made a fortune from the Studebaker purchase of his stock, but he was beginning to pay a price in terms of prestige and leadership when EMF had become incorporated within the Studebaker Corporation. Flanders and Fish had never been overly friendly toward each other, but both had recognized the need for each other in terms of advancing their respective companies. Now with Studebaker completely in the EMF driving seat, Fish may have begun to ask himself, "Was Flanders' presence still necessary?"

Part of that answer could have been read between the lines of an announcement that appeared in periodicals such as *The Motor World* for December 1911. According to this announcement, on Saturday, December 9, the Studebaker board had met, at which time J.M. Studebaker had tendered his resignation as president of the Studebaker Corporation. Frederick Fish was elected in his place. More ominously to Flanders' interests, Fish created a new position—general manager for the entire corporation, including the EMF automotive division. The man selected to fill the post was James Gunn of the New York firm of Gunn & Richards, Production Engineers. The caveat that Flanders would continue as Studebaker vice president and general manager of the automobile division may have been meant to keep him happy, but there could be no doubt that Flanders, a production genius in his own right, looked with disfavor on Gunn's appointment.[7-44] The end of 1911 left matters among Studebaker, EMF, and the Flanders Manufacturing Company, as well as those between Fish and Flanders, in an uneasy state. Into this sensitive scene entered Everitt and Metzger.

Chapter Eight

The Rebirth of Everitt and Metzger— Flanders, Where Are You?

The rumor spreading through the auto industry in early 1912 about Walter Flanders resigning from Studebaker Corporation (although he had more than two years remaining on his contract) was coupled with another rumor of equal weight—that Flanders soon would rejoin his old EMF partners, Everitt and Metzger.

It was no idle rumor. Everitt and Metzger indeed did start another automobile company, born out of the money they realized when they had sold their EMF stock to the Studebakers. However, Everitt staunchly denied the rumor of Flanders migrating to their factory. "E" and "M" were getting along quite well without "F," if he were to be believed.

Whether Everitt and Metzger had planned all along to form their own company before they arranged to sell their EMF stock was not immediately clear. However, industry watchers must have become suspicious when the two entrepreneurs convinced William Kelly, then the chief engineer of EMF, to resign with them.

Their suspicions would have been well founded. Less than six months later, on September 20, 1909, the Everitt/Metzger/Kelly team incorporated the Metzger Motor Car Company with the state of Michigan, listing it with a capital stock of $500,000, of which $300,000 already was paid in.

The company had been formed at least a month earlier, because in August the three men had talked Jacob Meier, a large trunk manufacturer, into selling them his cavernous factory on Dequindre and Milwaukee avenues and immediately began to fill it with machines related to auto production. The floor space of the building covered 156,000 square feet, and the structure sat on two and one-half acres of ground.

Officers (and the only stockholders) of the newly formed Metzger Motor Car Company were Barney Everitt, president; William Kelly, vice president and chief engineer; and Bill Metzger, secretary and treasurer. Management announced that once its factory was prepped for production, they would manufacture a runabout and a five-passenger car called the Everitt 30 to be sold at moderate prices. Prototypes of the two vehicles had been ready for some time and had become familiar sights on the streets of Detroit during the later months of summer when William Kelly took the cars on test runs. The officers also announced that they expected to manufacture approximately 5,000 automobiles for model year 1910.[8-1]

Bill Metzger, the consummate auto sales manager, again was up to his old tricks. As early as May 1909, after having sold his EMF stock to Studebaker, he began to spread the word among his dealer friends throughout the country that he soon would be back in the automobile business. Allegedly, with no more information than that, 200 to 300 established auto dealers sent their applications to him, indicating they would accept any type of automobile that Metzger would manufacture. What more encouragement could a man have for beginning his own business?

Was an Everitt 30 the Same as an EMF 30?

It is tempting to say that the Everitt 30 was an EMF 30 with a different nameplate because William Kelly designed both vehicles. However, this was not the case. The Everitt 30 conveyed a more rakish appearance. It had a four-inch longer wheelbase (110 inches vs. 106 inches), a "double-drop" frame that dipped three inches just forward of the dash to five and one-half inches at the rear before sweeping upward again, and an engine that was positioned much farther back than customary, so much so that the radiator was mounted behind the front axle.

Annette Kellerman, a famous actress in the early 1900s, driving her Everitt 30, circa 1910. (Courtesy of the Detroit Public Library, National Automotive History Collection)

In some specifics, the Everitt 30 and EMF 30 did bear a resemblance. Both had four-cylinder engines that produced 30 horsepower. Both engines had cast aluminum crankcases and used splash-type lubrication. However, the Everitt 30 had a bore and stroke that was four inches by four and three-quarter inches, versus four inches by four and one-half inches for the EMF 30, giving the Everitt 30 much better torque. Both vehicles had semi-elliptic front springs and full-elliptic rear springs, as well as an I-beam front axle.[8-2] However, the Everitt 30 weighed 2,200 pounds, 400 more than the EMF 30.

Cycle and Automobile Trade Journal, reporting on the Everitt 30 in its December 1909 issue, made much of the design simplicity of the vehicle:

> For instance, the upper crank case section, the four water cooled cylinders with integral water jackets, the inlet and

exhaust manifolds are cast in one complete integral unit. The simplicity of this construction is readily apparent. The cylinders are cast en-bloc, which form of construction is rapidly coming into favor on both sides of the water. Such a construction, a few years back, in fact not so far back, would have been virtually impossible, for not only must the designer and builder of the car keep abreast of the times, but the founder as well to whom reverts one phase of a successful product. This construction must be regarded as a mechanical masterpiece for it is by no means an experiment.

Reality Sets in for the Metzger Motor Car Company

Despite the accolades the press bestowed on the Everitt 30, the Metzger Motor Car Company fell considerably short of its announced goal of producing 5,000 vehicles in 1910, although the exact number is open for debate. One source states that first-year production was pegged at 2,500 units and already

This boxcar announces the delivery of Everitt 30 cars by train to Chicago. (Courtesy of the Detroit Public Library, National Automotive History Collection)

had been sold out before the first model was built.[8-3] If this were true, more customers purchased an Everitt 30 in 1910 than such established brands as Franklin, Oldsmobile, and Packard.

Probably much closer to reality was the copy that appeared in one of the Everitt 30 advertisements for 1911, which read:

> Then [after the prototype was completed and tested] the factory production began. Slowly, for quality was the object—not quantity. It was from the start impossible to supply the demand for a car like this. But altogether, in that year, 900 'Everitts' were placed in owners' hands.

It was quite possible that the Metzger Motor Car Company may have had twice or three times that number in orders, but chose or was forced to build fewer vehicles.

As time passed, all Metzger Motor Car Company advertising repeatedly stressed that the expressed intent of the company was to build a quality product. For example, in the same previously mentioned 1911 advertisement, the copy continued by stating:

> It was determined at the start to strictly limit factory production—a policy realized as the one means of maintaining quality... And that is why only 4,000 Everitts will be built this year [1911], in a factory double that capacity.

In all fairness to the Metzger Motor Car Company, any car company beginning from scratch in an empty building would have difficulty shifting into high-volume production during its first year under normal circumstances. Witness the problems that EMF had during its beginning stages.

The Metzger Motor Car Company Merges with the Hewitt Motor Company

Then there was the question of obtaining a license with the Association of Licensed Automobile Manufacturers (ALAM). Most successful manufacturers belonged to the association (which would be brought to its knees later by Ford), and Metzger felt that it was imperative for his young company to be a member. Metzger previously had been one of the directors of the ALAM. Therefore, he expected no difficulty in acquiring an ALAM license for the Metzger Motor Car Company. Unfortunately, times had changed. The ALAM had closed its ranks to newcomer companies. The Metzger Motor Car Company had no choice but to buy out an existing company owning such a license and take it over. For this, Metzger turned to the Hewitt Motor Company of New York and after much lobbying succeeded in convincing Edward R. Hewitt to merge his company with Metzger. Final papers were signed on December 31, 1910.[8-4]

Hewitt had been incorporated in March 1905 and was selling five different one-cylinder models and two eight-cylinder models when Metzger Motor Car Company absorbed it. Edward Hewitt may have been in the car business, but his heart was in trucks. Hewitt also made the heaviest truck in the United States at that time—a ten-ton behemoth that had found favor with breweries and coal-delivering companies. Hewitt's company also produced two-, three-, and six-ton trucks. When the company merged with Metzger, it introduced a light, one-ton van driven by a 17 horsepower, four-cylinder Everitt engine, probably designed by William Kelly.[8-5] When the merger was completed, Hewitt dropped its car business to focus entirely on trucks, and Edward Hewitt took his place on the Metzger board as one of its directors.

By the end of 1910, the Metzger Motor Car Company had undergone several changes. Capitalization had been doubled to $1 million, and the list of stockholders expanded from three to nineteen. However, Everitt, Metzger, Kelly, and Hewitt held 76 percent of the 7,200 shares of stock outstanding, according to the annual report of the company to the state of Michigan for 1910. Everitt and Metzger owned 3,666 1/3 shares between them; Hewitt owned 1,250 shares. Metzger acted as trustee for 800 shares that were set aside as future incentive for successful management personnel, a practice that had been initiated by Flanders at the founding of EMF in 1908. Looking at the annual report, one assuredly would predict success for the young company. Although it was more than a million dollars in debt, the company listed assets of $1.6 million, more than $200,000 of which was cash on hand and $392,000 in credits owed the company.

Metzger Product Line Expanded for 1911

At the January 1911 New York Auto Show, six different body styles for the Everitt 30 were unveiled, although the mechanicals were basically the same as the previous year. A five-passenger and four-passenger fore-door touring car were new additions, the fore-door quickly becoming the rage within automotive circles. It had the added attraction of locating all levers behind the front doors and inside the driver's compartment. What made an Everitt fore-door model stand out on the street was a molding that carried across the front and rear doors at the same height as the top of the hood, giving the car a long, low appearance from the side.

Also available were a light delivery wagon, the five-passenger touring model of the previous year, and a demi-tonneau. The tonneau (rear seat compartment) could be removed easily by unfastening two bolts.[8-6]

The Everitt 30 roadster, circa 1911. (Courtesy of the Detroit Public Library, National Automotive History Collection)

Hewitt Leaves Metzger Motor Car Company

In the spring of 1911, Edward Hewitt decided he would rather be back in the vehicle production business—his own. The Metzger Motor Car Company board accommodated him. At its May meeting, the board agreed to sell Hewitt the company's truck business, its New York property, and the Hewitt name for $100,000 in cash and $50,000 in common stock. In essence, Hewitt bought his own company back for the price of his stock in the Metzger Motor Car Company.

For some unknown reason, the new Hewitt Motor Company was not incorporated until December, probably because it took Edward Hewitt approximately six months to attract enough investment dollars to make his new (old) company viable. Its capital stock was listed at $1 million, and its product line consisted entirely of commercial vehicles.[8-7] A new factory and service station on West End Avenue and 64th Street in New York City were purchased in the fall of 1911. However, the factory was not fully equipped with production machinery until mid-January 1912, and the manufacturing of Hewitt trucks commenced then.[8-8] The newfound independence of the company was short-lived. A few months later, the International Motor Company (IMC) of Philadelphia, which included Mack Truck and the American Saurer Company, absorbed Hewitt. IMC, a holding company put together by J.P. Morgan & Company, then dropped the Hewitt truck line in 1913, although Edward Hewitt continued as an engineering consultant. IMC eventually adopted the name of Mack Trucks to eliminate confusion that the public was having between IMC and International Harvester.[8-9]

When in Doubt, Enlarge

Early in the fall, the Metzger Motor Car Company made plans for two large additions to the Dequindre plant—one being a four-story addition to the original complex that would extend it the length of almost two football fields. With these additions, the company announced that it would increase production to approximately 40 automobiles per day, placing a 10,000-car sales year within the realm of possibility. Such a sales year would be very nice. It could lift Metzger Motor Car Company into the ranks of the top half-dozen leading automobile manufacturing concerns of that day if those sales could be realized.[8-10]

The interior of the Metzger Motor Car Company factory in 1911. (Courtesy of the Detroit Public Library, National Automotive History Collection)

On October 5, 1911, at a special meeting of the Metzger Motor Car Company board, the directors made a curious move. They proposed that the company be dissolved and that all of its property and assets be placed in the hands of a trustee (meaning the current board of directors). The trustee would have the power to transfer said property and assets to a new company to be incorporated under the laws of the state of Michigan in exchange for capital stock that would be of equal value. The proposal was approved but nothing more came of it, at least for the time being. The Metzger Motor Car Company continued to do business as usual. Obviously, a shoe was meant to fall, but when and where? Speculation arose that if a next move was to be made by the Metzger Motor Car Company, it would involve Walter Flanders, despite the latter's seeming entrenchment within Studebaker management.

Launching the Everitt "Six"

The Metzger Motor Car Company then announced with substantial publicity that it was introducing a new six-cylinder model in keeping with its promise to produce affordable cars of quality, not quantity.

This was an extremely bold step by a company that was barely two years old. The number of major car companies offering or planning to offer a six-cylinder model at that time could be counted on a few fingers. Moreover, it required a substantial investment in the tools of production. No doubt the sale of Hewitt helped provide some of those funds, but without reincorporating to bring in more new capital, we can only surmise that the Metzger Motor Car Company was making money, and that money was being invested in the factory.

Evidently, offering the Everitt Six proved to be a good move. In one of its lengthy advertisements, the company stated[8-11]:

> ...when the first announcement of the wonderful 1912 Everitt was made, more than ten thousand interested persons responded, and, within six weeks, actual dealers' orders had been placed for five thousand of the new cars.

This would place the Metzger Motor Car Company in the same league as Hudson, Hupmobile, Mitchell, Oakland, and Reo, but whether 5,000 actually were built that year is unknown.

The Everitt "Six-48" came on a wheelbase of 127 inches in three body types: four- and five-passenger fore-door touring models and a two-passenger roadster. Dark blue was the only body color offered. A button on the dash operated a compressed air self-starter. (The company claimed it was the first to offer a positive self-starter as standard equipment for 1912.) Other essentials were a bloc-type long-stroke engine, double-drop frame, honeycomb radiator, and dual ignition. Featured was extensive use of expensive chrome-nickel steel that provided three times the strength of ordinary steel. Crankshafts, camshafts, connecting rods, each gear, the front axle—all were made of chrome-nickel. The price of the Six was $1,850.

Much emphasis was placed on the fact that the Everitt cars were not assembled from parts purchased from others, but were built of components manufactured entirely within the company's own factory, using a system of jigs and fixtures— 782 jigs to be exact. The company ad book stated:

> All Everitt cars are manufactured complete in one factory from the raw material to the finished product, thus doing away with the faults of inaccuracy, divided responsibility and uncertain workmanship necessarily common to all "assembled" cars.

These words sound remarkably familiar to those that Walter Flanders used when he talked about setting up production for EMF in 1908, including the use of jigs and fixtures. Startling, but not surprising.

Indeed, the Metzger Motor Car Company factory complex was quite advanced for its day. It contained a machine shop; an automatic screw machine department to produce screws, bushings, valves, etc., and to cut its own gears; a crankshaft department, which took in a rough forging and produced a finely balanced shaft; a grinding department for camshafts; and a hardening and heat treating department for front axle forgings and gears.

Along with the Everitt Six was introduced the Everitt "Four-36," a more powerful version of the previous Everitt 30. Body types offered were the same as those of the Six, but the Four-36 had a much shorter wheelbase of 115 inches. It too used chrome-nickel extensively, but it lacked a self-starter. Its selling price was $1,500. Also available was a basically carry-over version of the Everitt 30 from the previous year. At $1,250, it was the lowest-priced Everitt car ever offered by the company.

As 1911 came to a close, however, strong undercurrents of change for the Metzger Motor Car Company were in the air. What these changes would be were rife with speculation. Restlessness within the management ranks was evident—restlessness that seemed to develop whenever the name "Walter Flanders" was mentioned.

An Everitt 30 in the 1912 Buffalo Auto Club Run. (Courtesy of the Detroit Public Library, National Automotive History Collection)

Chapter Nine

Flanders Reunites with Everitt and Metzger

The year 1912 would be the last year in which an automobile manufacturer would believe that by manufacturing at least 10,000 units, it would be among the elite of the industry. In 1913, Henry Ford would begin feeding in the elements that would make up the first moving assembly line, and the old days of automobile manufacturing would be stood on end.

For Walter Flanders, the man who was a key component in setting up initial Ford Model T production, it was a wild year of ups and downs. For Everitt and Metzger, it was a year in which their own automobile company metamorphosed into a rather bizarre series of name changes and ownership.

It all began innocently. Early in January, Walter Flanders continued his negotiations to merge his Flanders Manufacturing Company with the Universal Motor Truck Company of Detroit. It was a strange move in light of his April 1911 letter to the Studebaker board, in which Flanders stated that he would "not invest in any other manufacturing enterprises during the term of my contract." The entire matter became moot when Universal eventually cancelled the discussions.

Nevertheless, it made more public the lack of harmony that had developed between Flanders and Studebaker since the previous April when Flanders had signed his latest contract to manage the latter's automobile operations. Talk had been circulating within the auto industry for the past three months that Flanders expected to bring matters between himself and Studebaker "to a head" after January 1.

During Christmas week of 1911, matters indeed appeared to have reached that sorry point. At that time, Flanders had met with James Gunn, the new overall general manager of Studebaker operations, and informed Gunn that he was severing his relations with the firm. Gunn refused to accept Flanders' resignation. He told Flanders that Studebaker would continue to pay him his salary and expected him to fulfill the letter of his contract. Flanders countered by threatening to take the matter to court to discover whether the contract (that he and his lawyer originally had drawn up) still was legally binding. Despite the fact that Gunn had refused Flanders' offer to resign, rumors were being whispered that Clarence Booth was being groomed to take Flanders' place as general manager of the automotive operations.

Flanders' relationship with Gunn appeared to have taken a turn for the better two weeks later when, on Tuesday, January 16, the board of directors of the Flanders Manufacturing Company met for its annual meeting. Flanders was re-elected president and Don McCord, the company general manager, became vice president. Conspicuous by their addition to the Flanders Manufacturing Company board were James Gunn and Scott Brown, who was the secretary and legal counsel for Studebaker.[9-1]

Going into 1912, the Flanders Manufacturing Company exhibited every sign of good physical health. In its annual report to the state of Michigan for 1911, the book value of the property stood at more than $3 million, of which $146,175 was cash on hand. Liabilities were listed at less than three-quarters of a million dollars. That was the good side. The bad side was that more than $2.5 million was tied up in materials and other kinds of tangible property.

Studebaker Rebuffs Flanders' Attempts to Resign

By February, it had become common knowledge that Walter Flanders had tendered his resignation as a vice president and general manager of the Studebaker Corporation. However, when broached on the subject, Flanders offered that the "if" and the "when" of this resignation were completely in the hands of the

Studebaker board of directors. As far as Flanders was concerned, the sooner the board released him from his contract, the better because he had an urgent need to devote all his energies to running the Flanders Manufacturing Company. Because Flanders had more than four years remaining on his present contract, there seemed to be little haste on the part of Studebaker to act quickly.[9-2]

As time passed without response from Studebaker on the Flanders matter, the rumor mills began to have a field day. One bit of gossip in particular caught the attention of the industry. This rumor centered on the activities of Paul Smith, a close friend of Flanders, who incidentally was the sales manager of the Studebaker Corporation. While visiting his home town of Indianapolis, Smith agreed to an interview in which he hinted that he was interested in building a six-cylinder car and that Flanders might be a part of that action. Talk now revolved around whether that six-cylinder car might not be the mysterious secret project that Flanders allegedly had underway with the Commercial Engineering Company of Detroit, another company that Flanders had purchased without fanfare the previous year.[9-3] Other bits of gossip linked Flanders with the Dodge brothers, who had broken off their relationship with the Ford Motor Company. Such speculation grew from Flanders being seen in company with the Dodges on several occasions.

The remainder of February and March 1912 passed without any comment from Studebaker regarding the release of Flanders from his contract. For his part, Flanders told those who would listen that he felt his resignation had taken effect because "I have been given very little to do around the Studebaker plants lately."[9-4] If anything, Flanders was upset because Studebaker informed him that the company had the right to use his name on its products even if he quit. Even more agonizing to Flanders' interests was that when the Studebaker board met in April for its annual meeting, most of the previous officers and directors were retained.[9-5] Flanders was returned to the offices of vice president and director, and no comments were made about his pending resignation. In effect, Flanders now was mired within a Studebaker "limbo."

A few blocks to the northeast of the Studebaker-EMF Piquette factory, the Metzger Motor Car Company was ready to make sensational news.

The Short and Happy Life of the Everitt Motor Car Company (a.k.a. Metzger)

On Tuesday, May 7, 1912, Paul Smith, erstwhile head of sales for Studebaker and close friend of Flanders, resigned from Studebaker. Two days later, Smith surfaced as head of sales for the Metzger Motor Car Company. More than that, Smith also assumed the position of first vice president and was rumored to have purchased a large number of shares in the company. Allegedly, Smith also brought new money into the fold in the form of Indianapolis businessmen. With him came his chief assistant, Frank Smith. Did this mean that Flanders soon would follow on Smith's heels? When approached, Bill Metzger refused to either confirm or deny the suggestion. However, company president Barney Everitt intimated that the acquisition of the two Smiths was only the beginning of a series of other tantalizing moves.[9-6]

What those moves might be were not revealed, but if they resembled the stories that had become the hot topics of conversation along the streets and newspapers of Detroit, something was brewing. For example, one rumor was that the Metzger Motor Car Company would merge with the Flanders Manufacturing Company and be reorganized into a new concern that would be capitalized at $3 million. Another was that a large new plant would be constructed on Woodward Avenue property that the Metzger Motor Car Company had purchased some time ago. Furthermore, the rumor claimed that the present Metzger Motor Car Company plant would build only the new six-cylinder Everitt cars, while the fours would be produced at the Flanders plants in Pontiac, Michigan. LeRoy Pelletier also was rumored to be on the way out as head of publicity at Studebaker, in order to assume a similar position at Metzger Motor Car Company. Then there was the item that had Flanders buying out the interests of both Everitt and Metzger in the Metzger Motor Car Company, which both men stoutly denied.[9-7] Detroiters had not had such fun in many years.

The Metzger Motor Car Company Evolves into the Everitt Motor Car Company

With all these rumors flying fast and furiously around Detroit, hardly anyone noticed that a new car company had been founded called the Everitt Motor

Car Company, which had the same officers as those in control of the Metzger Motor Car Company. Then, to the surprise of no one, the Metzger Motor Car Company was dissolved early in May. All property and assets were transferred to the new Everitt Motor Car Company in exchange for common stock amounting to $2.25 million plus $250,000 in preferred capital stock (the book value of Metzger Motor Car Company being $2.5 million). Par value of the stock was $10 per share. Inasmuch as the Metzger Motor Car Company originally was capitalized at only $500,000, this increase in its value over a three-year period rewarded its new/old owners quite well for their efforts. To be more specific, the Everitt stock was divided as follows: 70,000 shares to Paul Smith, 26,000 to Everitt, 26,000 to Metzger, 6,500 to William Kelly, 18,500 to assorted minority stockholders, and 28,000 to one trustee, plus 50,000 to a second trustee. Metzger received most of the first preferred stock—2,133 shares. Everitt controlled 283 shares, and William Kelly owned 84 shares. The remaining 50,000 shares were offered for sale.[9-8]

At its first board meeting, May 22, 1912, the seven directors of the Everitt Motor Car Company selected Barney Everitt as president, Paul Smith as vice president, Leslie Acton as treasurer, and Bill Metzger as treasurer. No surprises there!

At its June 5 meeting, the board made the rather strange pronouncement that the cost of manufacturing Everitt four-cylinder cars had reached a point where they no longer were profitable. The directors voted to drop them from production and offer only six-cylinder models for 1913. Evidently, more capital was needed, too, because they changed the terms of the sale of the Metzger property to the Everitt Motor Car Company: Everitt now was required to pay in a sum of $125,000 for his shares of the new company's common stock and Metzger $75,000.

For his part, Flanders persisted in refusing to make further comments about whether he would be joining his old friends. Then in August, that shoe finally dropped. The Studebaker Archives contains Flanders' letter of resignation from the Everitt-Metzger-Flanders Company dated August 7, 1912 to the EMF board, which the latter voted to accept. Although officially the automotive operation of Studebaker, EMF still had its own board, which would continue to exist for at least another five years after Flanders' departure. In the annual report to the state of Michigan for 1912, for example, Clement Studebaker, Jr. was listed as

EMF president, Frederick Fish as vice president, the lawyer Scott Brown as secretary, and a new entry—Albert Erskine—as treasurer. Studebaker Corporation owned all but three shares of the outstanding EMF stock.

Flanders finally was free to join his old teammates.

After Flanders left, Studebaker quickly removed any vestiges of the Flanders name from its passenger cars. New badges hurriedly were made, bearing the Studebaker title, and sent to Studebaker auto dealers to replace those that carried the titles "E-M-F 30" and "Flanders 20."[9-9] Walter Flanders now became no more than a memory within Studebaker records, and a contentious memory at that.

Be that as it may, the Studebaker Corporation stood in large debt to Flanders and EMF. Without his services and the plants and physical properties of EMF, there is every possibility that Studebaker would not have existed for another decade or two. Its wagon business was declining, and the number of electric cars and Studebaker-Garford cars that it assembled was so minuscule as to make survival with such products impossible over the long term. Fish and Flanders may not have been on good terms most of the time; however, Fish was a consummate businessman and would have been the first to recognize that Flanders and EMF provided Studebaker with a very successful and lasting entry into the automobile industry.

The Everitt Motor Car Company Becomes the Flanders Motor Car Company

Events involving Flanders in another venue now moved rapidly forward. At its August 21 board meeting, the Everitt directors voted to change the name of their company to the Flanders Motor Car Company and alter its certificate of incorporation to register this change. The capital stock of the company also was upped from $3 million to $3.74 million. The following week, another directors meeting officially noted that Walter Flanders now owned 1,762 and 1/3 shares of common stock of the company.

By mid-September, the new Flanders Motor Car Company announced that it would offer two six-cylinder cars for 1913. One would be called the "50-6" on a wheelbase of 127 inches and sell for $2,250. It essentially would be the Everitt Six with modifications such as a Gray & Davis electric starting and lighting system. Also, a gear-driven pump would replace the old vacuum splash system, and a pressure feed to the carburetor would take over for the previous gravity feed system. A new steering gear would be substituted as well.

The second six-cylinder car—a smaller, less powerful version of its sibling—would be advertised as the "40-6" and sell for $1,550. Its wheelbase would be 115 inches.[9-10]

At the September board meeting, Paul Smith moved that Walter Flanders be made the new general manager of the Flanders Motor Car Company. Of course, the board unanimously approved the motion.

Now the entire operating team was in place. It contained the familiar faces of Everitt, Metzger, Flanders, Kelly, and Pelletier, augmented by Paul Smith and two new members: Fred Hawes, reporting for duty after a decade as chief engineer at Cadillac; Bruce Ott, a well-known body designer; and Richard Miles, a leading metallurgist of that day. The Flanders Motor Car Company stood on the threshold of great things, reminiscent of four years previously when the Everitt-Metzger-Flanders Company strode into existence and became the largest employer in Detroit.

Would that same Flanders-driven magic work again? It might have, if the United States Motor Company had not intruded at that moment in time.

Enter the Ill-Fated United States Motor Company

The United States Motor Company was one of the two large auto conglomerates of its day, the other being General Motors. Founded with great expectations by Benjamin Briscoe in January 1910, the United States Motor Company was "no small potatoes." Capitalized at $16 million, it was made up of the Maxwell-Briscoe Motor Company, the Columbia Motor Car Company, the Alden-Sampson Manufacturing Company, the Dayton Motor Car Company, the Grey Motor Company, the Briscoe Manufacturing Company, and the Brush Runabout Company. The Maxwell-Briscoe Motor Company alone was the fourth largest auto producer in the nation.[9-11]

After a brilliant start, the United States Motor Company quickly came upon hard times. Similar to so many other new companies, it could not sell cars fast enough to pay the bills ran up by its satellite divisions. Unfortunately for Briscoe, by the time he learned that some of his subsidiary companies such as Stoddard-Dayton had grossly over-purchased on raw materials and were deeply in debt, it was too late. When creditors came knocking, they found a United States Motor Company that had most of its assets tied up in inventory and buildings, neither of which was easily converted into the quick cash necessary to satisfy the creditors. The United States Motor Company also had a real albatross to handle in June 1912 when it was faced with the payment of $750,000 in due notes.

The bankers, a nervous lot who owned the notes, met and decided to give the United States Motor Company a 90-day extension before payback. They also forced Briscoe to accept one of their members, W.E. Strong, as chairman of the board to oversee matters—in other words, to ensure that the company operated at a profit through the summer so that the bankers could recoup their loan money. By August, Ben Briscoe saw the handwriting on the wall. According to the August 11 issue of *The Detroit News*, he reportedly resigned from the corporation and sailed to Europe. What he left behind was a debt of $4.2 million to the banks that had extended their credit to the United States Motor Company, $2 million to merchants who had supplied materials and parts to the company, and $6 million in debentures (notes issued in lieu of cash to pay bills).[9-12]

United States Motor Company Taken to Court

On September 12, one day before the 90-day extension was to expire, the bankers were faced with a new challenge. The U.S. Motor suppliers, ostensibly fearing that the bankers would scoop up all available funds before they could be paid, filed a bill of complaint in the U.S. Court of Equity, alleging that U.S. Motor was insolvent and should be dealt with accordingly. The suppliers' case was led by the Brown & Sharp Manufacturing Company, to which U.S. Motor owed $70,000. In filing the bill of complaint, Brown & Sharp provided evidence that U.S. Motor had approximately $12 million in debts coming due by the end of October. Such an enormous sum could not possibly be repaid unless the company remained solvent; therefore, it should be completely reorganized unless the court had a better solution.

As was later learned, the bankers and the suppliers had come together at the beginning of summer with a working arrangement to recoup their investment. The 90-day extension the bankers had allowed on the notes that had come due in June was meant to give U.S. Motor time to sell all of its 1912 products before the public became aware of its troubles. Then, and only then, would someone such as Brown & Sharp act as a stalking horse to take U.S. Motor to court.

The unfortunate aspect of this entire episode was that the assets of U.S. Motor were valued at $15.3 million, far in excess of its debts. Moreover, its quick assets consisted of $9.25 million in the form of cash on hand, bills receivable, and cars either completed or in progress. These assets represented 75 percent of the amount due to the creditors, but the creditors feared that the remaining 25 percent would be too hard to come by over a short period of time and decided not to wait.

The court appointed W.E. Strong and Robert Walker to act as receivers and report at the end of October with recommendations as to what steps should next be taken.[9-13]

Flanders Asked to Take Over United States Motor Company

A reorganization committee hastily was formed to propose a new United States Motor Company organization that receivers then could recommend to the court. This involved the search for a successor to Ben Briscoe to head the company should the court accept the plan. First to be contacted was George W. Bennett of the Willys-Overland Company, but he refused the responsibility, as did several other important men in the industry. Finally, someone mentioned the name of Walter Flanders, despite the fact that Flanders had recently assumed management of the Flanders Motor Car Company.[9-14] Flanders was willing to listen. During the next month, he and the reorganization committee sparred back and forth over arrangements that would be satisfactory to both parties. Key to any deal, from Flanders' point of view, was that a reorganized United States Motor Company would purchase the Flanders Motor Car Company as part of his accepting their contract, thus combining both organizations into one.

Finally, on November 18, Judge Charles M. Hough of the U.S. District Court for the Southern District of New York heeded the suggestions of the receivers. He ordered the sale of the assets of the United States Motor Company and said he would entertain sealed bids for the property on January 8, 1913. Of course, it was understood that probably only one bid would be made, and it would come from those who were involved in the reorganization committee. The receivers immediately appointed Walter Flanders and William Maguire to manage the United States Motor Company until the January 8 sale date. Maguire, who had been the factory manager for the Ford Motor Company, was told that he would be made vice president and assistant general manager to Flanders after the sale was consummated and the reorganization had been completed.

Most of the dealings with Flanders had been semi-secret. However, in mid-November, LeRoy Pelletier, now advertising manager of Flanders Motor Car Company, announced that his boss, Walter Flanders, indeed would head the reorganized United States Motor Company after the sale of its property on January 8. Also, the Flanders Motor Car Company would be purchased by the new concern for a price of $3.75 million, of which $1 million would be in cash and $2.75 million in stock. Pelletier added that Flanders had made no plans to reduce the sales force employed by either company after the companies were combined.[9-15]

By the first week in December, large numbers of United States Motor Company management in sales, advertising, design, and engineering began to vacate the New York offices of the company. Flanders already had indicated that he would move its headquarters to Detroit. Most prominent of the leave-takers were J.D. Maxwell and Alfred Reeves. To those who remained, Pelletier said,[9-16]

> We want you to forget the past, because the old company was a failure and the name Flanders never has been associated with failure.

This remained to be seen, and Pelletier would not be around to chronicle the success or failure of the reorganized company.

Pelletier and Smith Resign

A few days after Christmas, the industry was completely surprised as both LeRoy Pelletier and Paul Smith tendered their resignations from the Flanders Motor Car Company, Pelletier as advertising manager and Smith as sales manager. At first, it was assumed that they had resigned from U.S. Motor rather than Flanders Motor Car Company. However, it was learned later that they never had been placed on the payroll of the former company.[9-17]

Rumor had it that Pelletier had drawn up a contract regarding his role in the Flanders/U.S. Motor reorganization, but Flanders had refused to sign it. Pelletier threatened to resign, and when he received no response, he did just that. On the other hand, Paul Smith claimed that the sales policies outlined by Flanders for the new company were not what he expected them to be (or at least his role in them) and also decided to resign.[9-18] It was a curious and abrupt end to a trusted relationship between the two men and Flanders—a relationship that had spanned many years and had led to many successes.

Almost unnoticed was news of the incorporation of a new automobile company called the Standard Motor Company, Inc. of New York on December 31. It should have attracted more notoriety because its capitalization was set at $31 million, a number well beyond the norm for a new auto concern of that era. However, reading the fine print, one would realize that the organizers of the new company were the same folks who were involved in the reorganization of the United States Motor Company, and the capitalization figure was similar to the one being bandied about for the reorganized version.[9-19]

Standard Motor Company Buys United States Motor Company: Flanders Remains in the Loop

On Wednesday, January 8, 1913, sealed bids on the United States Motor Company property were opened. There were two bids, both from the same source—the newly incorporated Standard Motor Company. The bids offered Judge Hough two alternatives for the purchase of United States Motor Company, both of which would easily satisfy the creditors. So it was left to the judge to decide which would be the better option. In studious attention of the proceedings was Walter Flanders, president-to-be of the Standard Motor

Company.[9-20] Judge Hough made his decision the next day, the sale was consummated, and the United States Motor Company creditors received a total of $8,475,000, a sum equal to the appraised value of all United States Motor Company properties.

Standard Motor Company Becomes Maxwell Motor Company: Flanders Made President

The Standard Motor Company now owned the United States Motor Company but in name only. Its legal experts already had discovered that corporations in several other states also were using the term "Standard." Therefore, the decision was made to allow the Standard Motor Company to die a graceful death and be replaced by another company called the Maxwell Motor Company, but under the same charter. The name may have changed, but the players remained the same, as did the original capitalization of $31 million. The latter would be increased to $37 million when the Flanders Motor Car Company was brought into the fold. It was expected that Walter Flanders soon would be named president and William Maguire vice president. Holdovers from the original United States Motor Company were Carl Tucker as treasurer and M.B. Anthony as comptroller.

It also was announced that two of the former United States Motor Company vehicle brands would be dropped—the Columbia and the Brush—but the Stoddard-Dayton and Maxwell marques would be continued. The Everitt Six car inherited from the Flanders Motor Car Company would assume the Maxwell name and continue in production.[9-21]

At the end of the month, after making the rounds of all the old United States Motor Company properties, Flanders stated that the high-priced cars of the new company would be made in the Columbia factory in Hartford, Connecticut. Its low-priced cars would be built in Dayton (home of the Courier, a low-cost Stoddard-Dayton), as well as in the former Brush plant in Detroit. To keep past buyers of United States Motor Company vehicles happy, replacement parts for those vehicles would be made in the Maxwell Motor Company plant at New Castle, Indiana. The ex-Maxwell factory in Tarrytown, New York, would be sold.[9-22] (One other United States Motor Company property, the Grey Motor Company of Detroit, already had been repurchased by its original owners.)

The next few months for the Maxwell Motor Company were spent in getting its act together. The old United States Motor Company was such a widespread operation that Flanders had difficulty sorting out the players and the plants from each other and deciding whom to keep, what to keep, and where to go from there when those things were established.

In early February 1913, Flanders announced that Maxwell would build only four models for the current season. Two of those models would be rebadged versions of the six-cylinder cars that were being marketed by the Flanders Motor Car Company: the larger Flanders 50-6, and the smaller Flanders 40-6. The seven-passenger 50 horsepower Six would sell for $2,350, and the 40 horsepower Six would sell for $1,550. Two four-cylinder models would complete the line: a 35 horsepower five-passenger touring car listing at $1,085, and a 25 horsepower touring car, its price yet to be determined.[9-23]

Maxwell Motor Company Finally Purchases Flanders Motor Car Company

The first meeting of the stockholders of the Maxwell Motor Company was not held until March 26, 1913, at which time a surprising discovery was made. Although Maxwell had taken over the management of the Flanders Motor Car Company, it did not own any of its property. In effect, Walter Flanders was the president of two companies: the Maxwell Motor Company and the Flanders Motor Car Company. That small matter soon was rectified as the stockholders present at their meeting on March 26, 1913 voted to acquire the Flanders Motor Car Company works for 20,000 shares of first preferred stock in Maxwell Motor Company; 20,000 shares of second preferred stock; and 20,000 shares of common stock. With par value of the Maxwell Motor Company stock being listed as $100 per share, the Flanders Motor Car Company stockholders thus stood to have a $6 million stake in the Maxwell Motor Company. These numbers added up to 20 percent of the latter's preferred stock and 18 percent of its common stock.

The stockholders of the Flanders Motor Car Company in turn met in a special session on April 4, 1913. They voted to accept the Maxwell tender. One result of this transaction was that Barney Everitt and William Metzger became two of the largest stockholders now on the Maxwell Motor Company rolls. However, they no longer were in any position of management authority. (Everitt did not formally resign as president of the Flanders Motor Car Company until May 2, and the directors of the company continued to meet until 1914.) The roles they shared with Flanders in creating a dynamic auto industry in Detroit had ended. The three men had made fortunes for themselves during the five years that they had been associated in business matters and were wealthy enough to retire in the prime of their lives.

Did their business relationship end with Flanders assuming control of Maxwell and the others fading into the background?

Not at all!

However, first we must return to mid-1912 and pick up the threads of another Walter Flanders venture, one that indicated betting on a Flanders enterprise did not always guarantee success.

Flanders Manufacturing Company Flounders... and Sinks

It is rare for one man to be the head of two different companies at the same time. It is even more unlikely to find a man who is the head of three companies simultaneously, but Walter Flanders was a risk-taker who reveled in the unlikely.

Consider that in August 1912, Flanders added the general manager's role in the Flanders (Metzger/Everitt) Motor Car Company to his list of responsibilities as president of the Flanders Manufacturing Company. Then, in early November, Flanders accepted the bankers' offer to manage a reorganized United States Motor Company. At this point, Flanders was in charge of three different companies.

Obviously, Flanders had spread himself too thin. Inevitably, one of the three companies would suffer from a lack of his attention. Which company would be the one to suffer was obvious. Because Flanders no longer held many shares of stock in the Flanders Manufacturing Company, its future success was not as important to him as it had been in the previous year. This soon began to show, especially because the Flanders Manufacturing Company was in trouble and needed a very active management to rescue it. That trouble could be traced to Flanders' own making and probably dated back to when he had resigned from Studebaker in August 1912. As long as he was running the EMF division of Studebaker, Flanders was able to direct a steady stream of business for parts and machining operations to his Flanders Manufacturing Company plants. As his active participation in production matters at Studebaker waned, Studebaker Corporation began to turn elsewhere for suppliers. Before this occurred, Flanders Manufacturing Company had been doing a million dollars a year in business with Studebaker. With its primary market now dissolving before its eyes, the Flanders Manufacturing Company desperately needed someone to develop new orders from other auto companies. Flanders was excellent at doing this, but he now had two other lucrative ventures to occupy the time he otherwise might have spent on replacing the Studebaker account.

Dragged Down by a Motorcycle and an Electric Car

Another problem dogging the company was the fate of its motorcycle and electric car projects that had been launched with such fanfare during 1912. Both were utter disappointments, despite the large amount of money being thrown at the development, manufacture, and sales of these products.

The motorcycle project had proved to be a disaster from the beginning. Ostensibly, it was the brainchild of LeRoy Pelletier, who convinced Flanders that it could be an excellent moneymaker. However, there seems to be every indication that the project was not thoroughly planned or evaluated. At first, the Flanders Manufacturing Company attempted to build a twin-cylinder motorcycle that would sell for the price of a single-cylinder model. As soon as its prototype was completed, a second program was initiated—to design a single-cylinder model that would be priced at $100. As it turned out, neither motorcycle could be sold at a profit. Both were commercial impossibilities. Having learned that fact of life one year and up to $300,000 later, the Flanders Manufacturing Company finally settled on a conventional motorcycle design but had been able to produce only about 2,500 by the fall of 1912, which was just as well because sales were not overwhelming.

The electric vehicle that had been introduced with such flourish during the summer of 1911 also proved to be a poor business venture. It was one thing to offer an electric car at $1,775, a price much lower than that of its competition. (Most electric vehicles such as the Baker or the Rauch and Lang sold for more than $2,000.) It was another thing to make an electric car that could be both cheap and reliable. The Flanders Electric proved neither to be reliable nor could it be made economically enough to bring in a profit at the $1,775 price. Initially, the company accepted approximately 3,000 orders; however, by the time it had worked out all the bugs of the vehicle system (if it ever did), the gloss had worn off sales. Eventually, Flanders had to raise the price another $500 in an attempt to make money from it. More than a year after its debut, only about 100 of the Flanders Electrics had been produced.[9-24] Both the electric vehicle and the three motorcycles had become enormous drains on the several millions that stockholders had invested in the Flanders Manufacturing Company.

As early as September 1912, Flanders had sent a message to the company stockholders that new money would be needed to carry on the business. He proposed that this money be raised by mortgage bonds or some other arrangement. For once, the stockholders rebelled. It was becoming quite clear to them, at least in regard to the Flanders Manufacturing Company, that Flanders' moneymaking magic may have deserted him. They refused his request.

The Final Demise of the Flanders Manufacturing Company

Word of the situation of the Flanders Manufacturing Company soon reached the creditors who, rightfully so, became alarmed. More than 69 strong, they came together on November 22 and selected seven of their members to act as a committee to determine how the company could best go about repaying them. The seven then selected three members to investigate more deeply what problems had to be overcome. These three reported back to the committee that approximately $300,000 of new money would be needed to keep the company solvent. Now it was the creditors' turn to refuse to invest further in the Flanders Manufacturing Company. In fact, several members were rather vociferous in demanding that the company declare bankruptcy. Their view was that the assets of the company were worth approximately $2 million, and, if sold, there might be enough money to pay all the creditors and perhaps some of the preferred stockholders. Holders of common stock, however, would lose their investments.

Most of the creditors held the view that it would be much better to have the court appoint a receiver, which then could place in control someone who might make the company viable. The Wagner Electric Company became their stalking horse. Wagner petitioned the federal court in Bay City, Michigan, to appoint such a receiver, alleging that although Flanders Manufacturing Company may have a future ahead of it, the company did not have sufficient funds at the moment to pay its bills. However, with good management, the company might pull through. The Detroit Trust Company was appointed as receiver.[9-25]

One of the first things that the receiver discovered was that the Flanders Motor Car Company had entered into an unusual contract with the Flanders Manufacturing Company the past August 1 to build numerous parts for the Flanders Six automobile. These parts would be built in the latter's Pontiac plant, using Flanders Manufacturing Company machinery and materials for which it was paying a fixed percentage above cost. The unusual aspect of the contract was that it stipulated (in a somewhat prophetic way) that the Flanders Motor Car Company had the right to continue building parts in the Pontiac plant even if the Flanders Manufacturing Company should go into receivership. As soon as this was discovered, the receiver cancelled the contract between the two companies but gave Flanders Motor Car Company new terms for leasing the property. These in turn were dropped when the three largest creditors of the Flanders Manufacturing Company complained to the court.[9-26]

By May 1913, the Detroit Trust Company had decided that the Flanders Manufacturing Company no longer was a viable enterprise, and the property would have to be sold to pay its creditors. The original projection—that such a sale would earn enough to satisfy the creditors and bring a partial return to stockholders—turned into gloom after a thorough appraisal was made of the company holdings. In essence, property that had been listed by the Flanders Manufacturing Company at a book value of $2,727,794 fell by 41 percent to $1,620,035 after the appraisal was completed. The largest loss—more than one-half million dollars—was attributed to so-called "fixed" assets (i.e., items such as good will, trade names, patents, deferred charges, contracts, and others) that the receiver deemed to be worthless. Current assets consisting of stores and work in progress were diminished by almost half to $292,849. The book value of the Chelsea plant fell by one-quarter million dollars. At the time the appraisals were made, the number of "proven" Flanders Manufacturing Company creditors stood at 696. They were owed $1,033,265.58. Assuming the receiver could sell everything at the appraised prices, those creditors could be repaid. (The ball bearing department already had been purchased by others, as were 39 electric cars and 142 motorcycles that were on hand when the receiver took possession of the company.)[9-27]

Everything now depended on the skill or good fortune of the Detroit Trust Company in disposing of the Flanders Manufacturing Company property. On June 10, a bid of $250,000 from the Harris Brothers & Co. of Chicago for the plant machinery was received and accepted. The Harris company also took a three-month option on the plant properties to purchase them for $175,000.[9-28] By this point, the expectations of the Detroit Trust Company had been reduced to being hopeful that it could bring a return of fifty cents on the dollar to the Flanders creditors. Obviously, this would mean that Flanders Manufacturing Company stockholders would be excluded from any return and would be left holding worthless stock certificates. Ultimately, they received nothing for their $2.6 million investment.

Thus, the Flanders Manufacturing Company wound down to an inglorious end. For the first time in his business career, Walter Flanders had failed his followers. Having sold most of his stock in Flanders Manufacturing Company to Studebaker earlier, Flanders personally was not as financially inconvenienced as his friends who joined him in this enterprise. Of note is the fact that Barney Everitt and Bill Metzger had avoided becoming any part of the Flanders Manufacturing fiasco.

Chapter Ten

"E" and "M" and "F" After 1913—

The Dance Continues

In the five years that had passed since 1908 when they introduced the EMF 30, Everitt, Metzger, and Flanders had made a profound mark on the rough and tumble days that characterized the early auto industry in Detroit. Rising from obscurity, the names of each of the three men had become household words within automotive circles, as well known to the public of that era as those of Henry Ford, Will Durant, R.E. Olds, or the Lelands. They also had made themselves millionaires in the process. By 1913, they had had a hand in the founding of six different automobile companies. One of those companies, Cadillac, continues to exist today. Another one of those companies, EMF, was the foundation on which the automobile operations of Studebaker Corporation were built. However, Everitt, Metzger, and Flanders were by no means as yet finished with their automotive endeavors.

Each now was prepared to go his separate way, although Everitt and Flanders would come together a decade later in one last grasp for automotive glory. The following chronicles the business ventures of the three men until the Great Depression, at which time they no longer were factors within the automobile industry.

The Metzger Years After EMF and the Flanders Motor Car Company

For Bill Metzger, the desire to remain a part of the auto world was too great for him to wander off into early retirement. Only 45 years old in 1913, Metzger was in the prime of his life, and he now had enough money to step back and indulge himself in other endeavors such as flying or becoming a leader in civic and business groups, although he continued to dabble in various projects involving the auto industry.

The Detroit Athletic Club

In December 1912, Metzger and 15 other familiar figures such as Henry B. Joy, Hugh Chalmers, Roy Chapin, and Frank Stearns met for lunch in the Pontchartrain Hotel, from which meeting came the formation of the Detroit Athletic Club (DAC). These men and other auto industry barons raised $2 million to erect a new building in downtown Detroit at which auto industry gentry could dine, relax, and enjoy each other's company. Completed in 1915, the DAC has become one of the most familiar institutions in Detroit. Its membership roster during the past 80 years has contained the names of the leading auto industry figures, past and present. Its meeting rooms are used today for elaborate automotive press conferences, especially as a venue for auto suppliers to introduce new products.[10-1] Bill Metzger had the honor of serving as DAC president in 1921. At that time, its membership stood at 3,250, a six-year growth that was truly astounding. Most members were the industrial leaders of Detroit.

Auto Parts Manufacturing Company

Metzger's first automotive attachment away from either Everitt or Flanders was as vice president of the firm of an old friend, Alfred Owen Dunk, who owned the Auto Parts Manufacturing Company, less than a half mile east of the EMF plant.

For Metzger, it was a step into the past. The Auto Parts Manufacturing Company (Auto Parts) originally had been set up by Dunk in 1908 at the request of Walter Flanders to build parts for the Wayne and Northern cars after these two companies were merged into EMF. Auto Parts later took over all the stock and blueprints of the De Luxe Motor Car Company in March 1910 when it had been purchased by EMF. On the De Luxe property, EMF then built the Flanders 20. Again, it was a good-faith gesture on the part of Flanders to provide those who had purchased De Luxe automobiles with the parts necessary to keep the cars in running order.

Dunk left Auto Parts in 1915, at which time Metzger replaced him as president, a position he continued to hold until 1924 when Auto Parts merged with the Puritan Machine Company (another parts company owned by Dunk) to become the Puritan Autoparts Company. Dunk probably left Auto Parts because he had acquired a new interest, the Briggs-Detroiter Company, which had just gone into receivership. Dunk negotiated its purchase in 1915, then changed the name of that company to the Detroit Motor Car Company and resumed production. (The Detroiter car lasted two more years.)[10-2]

Hail, Columbia

On January 15, 1916, Metzger joined with J.G. Bayerline, T.A. Bollinger, A.T. O'Connor, and Walter Daly in organizing the Columbia Motor Car Company. Bayerline, Daly, and Bollinger had just quit the King Motor Car Company (without any advance warning), where they were president, sales manager, and plant manager, respectively. They formed their own company to build a more expensive automobile that would have much higher prestige—with Metzger's help.

Columbia was capitalized for $500,000, but only slightly more than $25,000 of the common stock was paid for in cash. Approximately $175,000 in stock was represented by the property involved in the company operations which were conducted in a leased facility at Jefferson Avenue and Bellevue Street, a block away from where the original Olds factory in Detroit had been located. Bayerline was elected president, Bollinger vice president, and O'Connor secretary/treasurer. These three men plus Daly and Metzger owned 96 percent of the common stock of the firm and with J.S. Mohrhardt formed its board of directors.[10-3]

The first Columbia was introduced at the New York Auto Show in January 1917 after a year of preparation under chief engineer Ray Long. It had been necessary to purchase the assets of the American Electric Car Company in November to acquire floor space in the show. American Electric, which had just gone bankrupt, owned a license in the National Automobile Chamber of Commerce. Without such a license (and no more were being issued), exhibiting a product in national auto shows was denied.[10-4]

Makeup of the Columbia

The Columbia was purely an assembled car. The factory made nothing of its own. However, its management was discriminating in its choice of suppliers and produced a quality automobile. A six-cylinder, L-head Continental engine that displaced 225 cubic inches and produced 38 horsepower was an excellent choice for the Columbia Six. Stromberg manufactured its float-feed carburetor. A Stewart vacuum system provided the gasoline feed from a rear-located 17-gallon tank. Timken was the source for front and rear axles. A Ward-Leonard two-unit starting and lighting system was accompanied by a Bendix drive. The ignition came from Atwater-Kent, and the storage battery from Willard. Warner supplied the steering gear and three-speed manual transmission. Borg & Beck made the 10-inch single dry plate clutch. The frame came from the Detroit Pressed Steel Company and the springs from the Detroit Steel Products Company. Exclusive to the Columbia was a special radiator from Harrison. It featured a thermostat above the fan, which automatically opened a set of shutters to regulate cooling as engine temperature increased. Body color was a choice of blue or maroon. Wheelbase was 115 inches. Only a touring car was available initially, and no one could quibble with its asking price of $1,145. The company sold 800 cars in 1917, the first year of production.

In 1918, Metzger replaced Bollinger as vice president. Columbia now offered two other models—a sport touring and a sedan. Prices ranged from $1,350 to $1,995.

Sales continued to climb. For model year 1920, Columbia produced 4,806 automobiles, which was more (in some cases, much more) than several well-known makes such as Moon, Marmon, Pierce-Arrow, Peerless, or Stutz.[10-5] It is reasonable to assume that the net income from the sales of Columbia

"E" and "M" and "F" After 1913

The Columbia Six as it was advertised for 1918. (Courtesy of the Detroit Public Library, National Automotive History Collection)

cars that year was approximately $1.5 million versus Columbia's first year of existence when it was doubtful if the company netted more than a quarter million dollars.

Columbia Reorganizes

Flushed with their success, Columbia directors reorganized the company in 1920, increasing its capital stock to $4 million simply by declaring a huge stock dividend of 700 percent, which was paid in the form of shares of Columbia common stock. The next year, capitalization was raised to $5 million to be paid for by subscription. Then it was increased to $6 million.

Early in December 1922, *The Motor World* carried the news that Columbia was planning to produce 27,000 automobiles during 1923. On the face of it, this was a rather ambitious forecast for a company that had made less than a quarter that number three years previously. There would be three new body types, giving the customer a choice between four models of the Light Six and two of the more expensive Elite models; however, there would be little or no change in the engine and chassis. In late June, Columbia stockholders voted to reduce the capitalization of the company from $6 million to $3 million.

Columbia sales peaked in 1923. At its annual meeting at the beginning of July, Bayerline, the Columbia president, reported that the company had more than 5,000 vehicles ready for immediate delivery, and enough orders to keep the company at maximum production for the next six months. The newly introduced Light Six made up 23,700 of these orders, the company claimed.[10-6] What happened to those projected orders remained to be seen, for company production that year tabled out at 6,000.[10-7]

Regardless, Columbia could not withstand its own prosperity. It had been assembling automobiles in a leased building when an opportunity presented itself during the summer of 1923 to purchase a permanent residence—the still new factory complex of the Liberty Motor Car Company that had gone into receivership.

Columbia Motor Car Company Absorbs the Liberty Motor Car Company

Although its founding in 1916 was even more humble than that of the Columbia Motor Car Company, the Liberty Motor Car Company had shown spectacular growth in the first five years of its existence. Percy Owen, who once had been a salesman for the Winton Motor Car Company, had organized it in February 1916 with a capitalization of only $400,000. Joining Owen were R.E. Cole and H.M. Wirth. All three men came from the Saxon Motor Car Corporation, which had risen to eighth place in sales. Saxon was a maker of low-priced cars, none of which sold for more than $1,000. The Liberty was a decided step upward in class, its price ranging from $1,095 to $2,350.

After a first-year output of 733 cars, production increased to 1,900 in 1917, then to 4,414 by 1920, with a spectacular jump to 11,000 in 1921. The future appeared bright enough for the company to build a huge new plant on Charlevoix Avenue, not far from the Hudson and Chalmers auto companies. A picturesque administration building styled after historic Independence Hall in Philadelphia fronted it. As quickly as the Liberty rose, even more quickly did it fall. The attempted expansion was too rapid, and the company no longer could pay its bills. A committee of creditors took over its management in January 1922, looking for a way to refinance it or find a buyer to reorganize it. After 12 months of searching within a depressed market, the creditors gave up hope and asked the courts to liquidate the company.[10-8] The Security Trust Company was appointed as receiver on January 3, 1923, and in May advertised the property for sale at $1.175 million.[10-9] At this point, the Columbia Motor Car Company stepped in and by September had taken over the 24-acre Liberty site. Columbia immediately began to move into its new quarters, this being its first permanent residence since the company was organized.

The Liberty Acquisition Drags Down Columbia

The price of the new residence soon became as much of a burden to Columbia Motor Car Company as it had been to Liberty Motor Car Company. One year after purchasing the latter's property, Columbia also was unable to pay its creditors and went into bankruptcy. Again, the Security Trust Company was appointed as receiver. An appraisal of the property determined Columbia assets to be $997,455.16 offset by liabilities of $863,559.64, of which almost half were made up of mortgage, back rentals, and taxes on the property taken over from Liberty. Although the receiver hoped to find someone to invest new money so that the company could be reorganized, the firm of Winternitz & Tauber, a Chicago-based auctioneer and liquidator of defunct companies, made a bid of $112,500 for the Columbia machinery, plant equipment, and personal property. The Security Trust Company felt obliged to accept the bid. The plant then was put up for public auction.[10-10] The following year, 1925, the Edward G. Budd Manufacturing Company purchased it and continues to maintain it in pristine condition to this day. Although Columbia creditors realized some gain, the stockholders, including vice president Bill Metzger, realized nothing for their original investment.

Enter the Wills Sainte Claire

Even while serving as vice president of Columbia, Bill Metzger continued to dabble in other automotive activities. In 1921, he joined forces with William Hurlburt to form the Wills Sainte Claire Company, acting as the Michigan region sales distributor for the newly formed C.H. Wills & Company. Metzger was the president, and Hurlburt was the vice president. Hurlburt was once owner of the Hurlburt Motor Truck Company of New York in 1912. It had done quite well until the government stopped buying trucks after World War I and unloaded those remaining on hand onto the commercial market. Hurlburt then became assistant sales director for the Maxwell Motor Company.[10-11]

The Wills Sainte Claire car was Childe Harold Wills' initial venture into building the perfect car after leaving the Ford Motor Company, to which he had contributed much success. His car was pricey, ranging from $2,875 for an open touring car to $4,775 for a limousine.

Wills' penchant for frequently stopping the production line to introduce changes, aside from the fact that he picked a time of stark recession to introduce a new automobile, led his company into receivership in November 1922 with an accumulated a debt of $8 million. Wills stubbornly repurchased it from the receiver for $750,000 with the help of Kidder, Peabody and Company and reorganized it, only to see it fade into liquidation in late 1926. Approximately

"E" and "M" and "F" After 1913

A 1924 magazine advertisement for the Wills Sainte Claire. *(Courtesy of the Detroit Public Library, National Automotive History Collection)*

12,000 of the Wills Sainte Claire models were built over the five-year life of the company, and Wills went $4 million into debt, a debt which he eventually repaid.[10-12]

Two years after the demise of the Columbia, Metzger became a director of the Federal Motor Truck Company, a leading Detroit truck manufacturer founded in 1910. Federal offered models from one and one-quarter ton to five and one-half tons when Metzger came onboard. Shortly thereafter, he was appointed to the vice presidency of that company. Federal had established a reputation for the conservative styling of its trucks, which may have contributed to the longevity of the company when others featuring more avant-garde designs found them unacceptable.[10-13] The Federal Motor Truck Company association actually brought to a close Bill Metzger's propensity for becoming involved in the rise and fall of companies that engaged in vehicle transportation.

Metzger's Civic and Industry-Related Activities

In addition to his business ventures, Metzger made his presence felt within the auto industry and the city of Detroit in so many other ways that we can only conclude that such activities were his primary hobbies. For example, in October 1918, he was elected treasurer of the newly organized Michigan State Council of Motor Clubs, which had a charter to lobby local and federal government for better roads and road marking.

Then, two months before the dissolution of the Columbia Motor Car Company, Metzger was appointed to the executive committee of the American Automobile Association (AAA), the pioneer of all automobile associations. It had recently combined with its main competitor among such national groups, the National Motorists' Association. Metzger had been one of the prime movers behind the merger.[10-14].

In March 1924, Metzger also was elected president of the Board of Fire Commissioners for the city of Detroit. It simply was another civic duty to combine with his being a city supervisor, one of Wayne County's under sheriffs, president of the new park board in the county, chairman of the good roads committee of the Board of Commerce, and chairman of the transport committee of the State Highway Association.

In the mid-1920s, Metzger was a member of the board of directors of the National Automobile Chamber of Commerce and served as the chairman of its traffic and insurance committees. His fellow directors were such illustrious auto men as Roy Chapin, Alvan Macauley, Windsor White, F.J. Haynes, R.E. Olds, and John Willys.

In June 1926, while acting as vice president of the Detroit Automobile Club, Metzger was selected to chair the prestigious Conference on Uniform Traffic Laws to be held at the General Motors Building on September 8. This led to his appointment the following year as chairman of the National Conference on Municipal Traffic Codes by the Secretary of Commerce at that time, Herbert Hoover. Its charter was to draw up a set of uniform regulations for the control of vehicle and pedestrian traffic in urban areas.

There simply was no end to Metzger's volunteer activities.[10-15]

A Passion for Aircraft Companies and Flying

From watching the Wright brothers launch their early flights to bringing one of the first Wright planes to Detroit, forming the Aero Club of America, drifting around the countryside in a balloon, or flying one of his own planes, Bill Metzger had an unabashed enthusiasm for air travel. Just as he had made his fortune within the auto industry, he also was attracted to the aircraft industry and did what he could to forward its progress.

After the demise of the Columbia Motor Car Company, Metzger began to devote much more of his time to the aircraft industry. That industry was sorely in need of promoters, inasmuch as it had fallen into deep decay after World War I. Only 60 planes were built in 1924 for civilian use, for example. In 1925, however, the government gave commercial aviation a real boost when it began awarding mail contracts to private companies.[10-16]

Metzger immediately became part of the action. That same year, he was a member of the executive committee of the newly formed National Air Transport Company, a $10 million venture to carry mail and passengers between Detroit and Chicago. He also was elected to the presidency of the Detroit Aviation Society.

In January 1926, Metzger helped Eddie Stinson form the Stinson Aircraft Company. Stinson began a small plane revolution by placing the pilot inside an enclosed cabin and providing his plane with a self-starter and brakes. The company became the largest Michigan manufacturer of airplanes and endured until 1946.

In 1928, Metzger organized the first All-American Aircraft Show in Detroit, of which he was vice chairman.

In February 1929, Metzger and a number of auto leaders such as Ross Judson (president of Continental Motors), Dubois Younger (president of the Hupp Motor Car Company), Howard Coffin (vice president of the Hudson Motor Car Company), and other leading Detroit businessmen organized the Cadillac Aircraft Corporation. They leased a factory and flying field in Northville, Michigan. Metzger was installed as president and H.G. McCarroll as vice president. McCarroll had designed a twin-motor amphibian, the only one of its kind in the industry, which Cadillac intended to sell at half the price of a single-engine amphibian. Its power came from a pair of 110 horsepower Kinner motors that produced a top flying speed slightly over 100 miles per hour. Unique to the new aircraft was a patented mechanism that retracted the wheels into the wings when the pilot intended to land on water. It was so unique at the time that the entire industry followed its progress closely.[10-17] However, the effects of the Great Depression arrested that progress.

Metzger was only 61 years old at the onsct of the Great Depression. He was a wealthy man in the prime of his life—a man who had given much of himself to the city in which he lived, as well as to both the business and private sides of the automobile and aircraft industries. He was one of a small circle of men whose name had become a household word within the Detroit business community during the first three decades of the twentieth century.

The Flanders Years After EMF and Flanders Motor

The Surprising Success of the Maxwell Motor Company

Out of the ashes of the United States Motor Company, Walter Flanders slowly resurrected the Maxwell Motor Company in 1913. EMF was a thing of the past, as was the Flanders Motor Car Company, which now was merged into Maxwell. Production at all United States Motor Company plants had been stopped the previous year after it went into receivership. Flanders began writing off most of them as "not being adaptable to automobile manufacturing" when he became president. It was a strange comment for Flanders to make, inasmuch as the manufacture of automobiles was the primary business of the United States Motor Company.

With the finalization of the purchase of the Flanders Motor Car Company by the Maxwell Motor Company in April 1913, Flanders took the final steps to eradicate any residue of the United States Motor Company. The latter's plants in Tarrytown, Providence, Auburn, Newark, and Hartford and one plant in Dayton were assessed for their value preparatory to putting them on the block. The Newcastle, Indiana, plant was omitted from this picture. Instead, it was outfitted with machinery to make replacement parts for past United States Motor Company vehicles, as well as to provide forgings for all future Maxwell cars. In due time, the Newcastle plant would become one of three main Maxwell Motor Company car manufacturing facilities.

Flanders' First Maxwell

Three new Maxwell models for 1914 were announced. Most expensive would be the Model 50-6, a six-cylinder, 41 horsepower touring car on a 130-inch wheelbase, selling for $1,975. The smallest would be the Model 25-4, powered by a four-cylinder, 21 horsepower engine. It had a 103-inch wheelbase and was priced at $750. Between the two was the Model 35-4 on a 110-inch wheelbase, also carrying a four-cylinder engine but one rated at 26 horsepower. It would sell for $1,225.[10-18] Flanders said that the 50-6 would be built in the previous Flanders Motor Car Company plant on Milwaukee Avenue in Detroit, the 25-4 at the ex-Brush and Stoddard-Dayton plants also in Detroit, and the 35-4 in Dayton.[10-19]

On July 22, 1913, Flanders presented the Maxwell Motor Company executive committee with a production schedule for the three models, which it then approved, especially when Flanders indicated some of the potential profit margins that were involved if a certain level of manufacture could be maintained. For example, a build of 500 to 700 of the Model 35-4's each month would bring a profit of $415 per car; a build of 1,200 to 1,400 of the Model 25-4's per month would bring a profit of $224 per vehicle. Once these were sold, the Maxwell Motor Company could realize a net profit from these two models of $5.5 million at the low end to as much as $7.5 million at the high end. As events later were to prove, Maxwell did not achieve a total output equal to the low end, but it did come within 90 percent of that number—enough to make Flanders' forecast a success.[10-20]

In June 1913, the Maxwell Motor Company was jarred by the abrupt resignation of William Maguire, vice president and assistant general manager to Flanders (and possibly considered his heir apparent). Maguire's only comments in leaving were,[10-21]

> You can best describe the situation by saying that I was not and am not in accord with Mr. Flanders' principles.

Maguire did not elaborate, but some friction clearly had arisen between him and Flanders, perhaps relative to Flanders favoring "his own men" in the organization. Maguire did not resign from the Maxwell Motor Company board of directors, however. Regardless, his resignation had little effect on the progress of the company under Flanders.

Sale of United States Motor Company Plants in Full Swing

By the fall of 1913, Flanders began to find buyers for the former United States Motor Company properties. The Dayton Motor Car Company, builder of the highly regarded Stoddard-Dayton car, was one of the first to go. In August, the Maxwell Motor Company board accepted a bid of only $4,000 for the assets, then, in September, placed the Hartford, Connecticut, plant of the venerable Columbia car on the market for an asking price of $400,000.

In May 1914, the Maxwell board approved the sale of the Milwaukee Avenue plant, original home of the Everitt car, for $200,000. Production of the Maxwell six-cylinder models at the plant had stopped when Flanders decided to focus all of Maxwell's energies on four-cylinder models. The board also accepted the sale of the Briscoe Manufacturing Company for $100,000. This sale ended an era because that company had brought Ben and Frank Briscoe into the automobile business via the manufacture of radiators and other sheet metal items at the turn of the century.

The most newsworthy of the plant sales was that of the old Maxwell-Briscoe Motor Car Company plant in Tarrytown, New York, on the banks of the Hudson River. The Chevrolet Motor Company picked up that plant for $267,000 and would use it for the production of Chevrolet cars. The Tarrytown plant had been the centerpiece of the Maxwell-Briscoe Motor Car Company operations under the United States Motor Company banner.[10-22]

Maxwell Motor Company Begins to Show Excellent Profits

In October 1914, Flanders reported that the Maxwell Motor Company had realized net earnings of $1.43 million and now had a net working capital of approximately $6 million and cash on hand to the extent of $1.79 million.[10-23] Consider that less than two years earlier, Flanders had taken control of a reorganized and renamed ex-United States Motor Company teetering on the edge of bankruptcy and now had it pushing Studebaker for fourth place in the industry. It was a brilliant piece of work.

By the 1915 model year, Flanders had pared the Maxwell model offerings to one—the four-cylinder, 21 horsepower Model 25 on a 103-inch wheelbase. It was scarcely changed from the previous year, but its price in touring car form had been reduced from $750 to $695. Three new body styles were introduced: a roadster, a cabriolet, and a town car, the latter selling for $920. These prices compared favorably with all Ford Model T's, except those on the low end of the scale. Maxwell Motor Company advertisements touted the 1915 model as "one of the greatest hill climbers in the world." The ads claimed that a Maxwell could step up from 4 miles per hour to 50 miles per hour in a "comparatively few yards" in low gear. It also featured an adjustable front seat, "an improvement not found in any other automobile."[10-24]

The decision to reduce Maxwell offerings to one model in 1915 proved to be most providential. Production soared to 44,000 units, giving Flanders the satisfaction of exceeding the output of the Studebaker Corporation and of boosting Maxwell Motor Company production to within an eyelash of Dodge and a close striking distance of Buick.

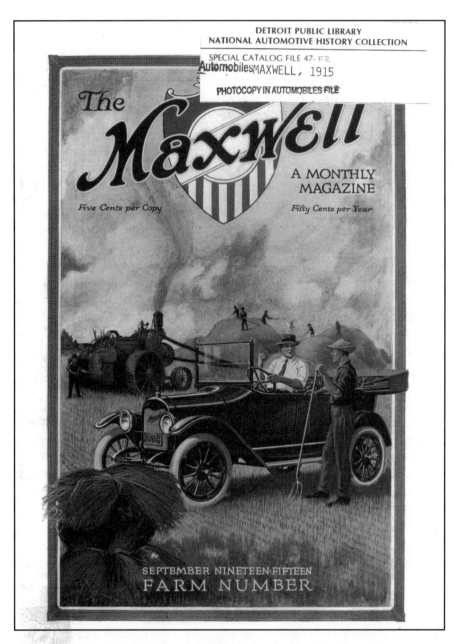

This 1915 Maxwell touring car was shown on the cover of The Maxwell, *a monthly magazine. (Courtesy of the Detroit Public Library, National Automotive History Collection)*

"E" and "M" and "F" After 1913

A 1915 advertisement for the Maxwell Motor Company.

The year was financially rewarding for Flanders as well. His contract called for an incentive payment of four percent of the earnings of the Maxwell Motor Company. Thus, nobody was surprised when the executive board approved the payment of $100,467.18 to Flanders during its September 1915 meeting for his services of the preceding sales year. This was an enormous sum of money in a day when income taxes did not yet exist and the average wage earner in an auto factory brought home much less than $2,000 a year.

In 1916, Maxwell added a sedan to its lineup; otherwise, the model array was the same. Changes were minimal, the most noticeable being an enclosed flywheel and clutch assembly and a new instrument panel. The prices of all models again were reduced, with the sedan at the low end selling for $635, actually lower in price than the comparable Model T body style. On July 1, 1916, Flanders dropped the Maxwell prices even lower, the five-passenger touring car now selling for $595 and the roadster for $580.[10-25] In typical Pelletier publicity hype, Maxwell set various fuel economy records during the year. Likewise, to prove the durability of the Maxwell power plant, a Maxwell was run to a nonstop record of 22,000 miles on the West Coast during January 1916.[10-26] Production rose to approximately 69,000 units. Again, Flanders had outproduced his rival, Frederick Fish at Studebaker.

The year 1917 proved to be Maxwell's greatest year. Production peaked at 75,000 units—almost double that of Studebaker. To Flanders, it must have been sweet revenge for being pushed aside during his last year at Studebaker. To the bankers who had asked Flanders to reorganize United States Motor Company five years previously, the Maxwell Motor Company balance sheet must have been a joy to behold. As usual, Flanders was paid well for his efforts. On September 25, 1917, the Maxwell board approved the payment of $289,878.79 to Flanders per his contract agreement for five percent of Maxwell earnings for fiscal year 1916.

Chalmers Plants Leased

In September 1917, the Maxwell Motor Company entered into an agreement with the Chalmers Motor Corporation. Chalmers was not using all of its production capacity, and Maxwell needed more manufacturing space. Flanders entered into what later would become an ill-fated agreement in which Maxwell leased the entire Chalmers Motors operation and ran it, building both Maxwell and Chalmers cars on Chalmers properties. Hugh Chalmers, the company president, moved up as board chairman to remove himself from the day-to-day activities of the company.

Flanders Retires from Maxwell

Flanders' five-year contract as president and general manager of the Maxwell Motor Company was coming to a close. To clear the way for a predecessor, he announced in early December 1917 that he had resigned from the presidency of the company. Actually, he was leaving the administrative management of the company to his successor, W. Ledyard Mitchell. Flanders continued to hold the position of general manager. Then, on the following month, the Maxwell directors elected Flanders as board chairman. News of Flanders' pending retirement brought accolades from everyone within the industry and the automotive press regarding his accomplishments at Maxwell. *Automobile Topics* probably summed it up the best when it wrote[10-27]:

> The Flanders Administration of the Maxwell company has been one of the real and noteworthy successes of the automobile business. In it Flanders not only conclusively demonstrated what was already known to those who knew him, namely, that he is a great manufacturer, but he proved equally forcibly that he is just as great as an organizer, merchant and the financial master of his business. When he took the sadly scattered remnants that were once the United States Motor Co., the task seemed a Herculean one. Flanders performed it in the simplest possible manner, cutting out not only the various models and kinds of cars, but also disposing of the plants themselves and unifying the Maxwell structure, concentrating it around the production of a car whose exact field and market he had in mind. The growth of the business had been nothing short of marvelous, and today it is rated among the most solid of the industry.

Flanders gradually withdrew from the daily operations of the Maxwell Motor Company to ensure that his final departure would not be too disruptive. He walked out of its doors into retirement on July 31, 1917, having elevated Maxwell into a corporation that was worth $37 million and was one of the six largest automobile manufacturers in the nation.

The management that controlled the Maxwell fortunes after Flanders' departure was unable to maintain the high level of quality that Maxwell cars enjoyed under Flanders' leadership. The company also drifted into another direction of business, one that looked toward a profit via lower volume and higher prices for its products. Production never again would reach the peaks that the Maxwell Motor Company achieved under Flanders, and the company gradually fell into desperate straits. Another automotive giant, Walter P. Chrysler, would rescue the company in much the same manner as Flanders had done when converting the United States Motor Company into the Maxwell Motor Company, reorganizing it and changing its name until it became a new force within the industry.

The Everitt Years After EMF and the Flanders Motor Car Company

Back in the Body Business

The sabbatical that Everitt took after winding down the affairs of the Flanders Motor Car Company in mid-1913 was short-lived. The *Detroit City Directory* for 1913 listed Everitt as the president of the Beamer & Bryant Building and Realty Company and the B.F. Everitt Company, a machine shop.

In 1915, Everitt formed the Everitt Brothers Company, which he incorporated the following year. It was a return to his roots, the trade that initially brought him so much success, wherein he did automobile body painting and trimming, and manufactured tops. His brother Roland was vice president; his brother Gordon was secretary and treasurer.[10-28]

In hiring his brothers, Barney Everitt was not being paternalistic. When Walter O. Briggs had bought Everitt's first body company in 1909 and formed his own company around it, he had made Roland his vice president. Thus, Roland knew the body trimming business quite well. The company was located at 669–688 Mack Avenue in Detroit. A second factory was acquired in 1916 at 1250 Jefferson Avenue.

"E" and "M" and "F" After 1913

Barney Everitt, taking a break from his business for a boating excursion on Lake St. Claire. (Courtesy of the Detroit Public Library, National Automotive History Collection)

The following year, Everitt was asked by the failing Springfield Metal Body Company to take over its affairs as president; however, before that took place, Springfield's creditors forced the company into court and Everitt was appointed as its receiver.[10-29]

Everitt and the Rickenbacker Experiment

Shortly after World War I, Captain Eddie Rickenbacker, America's most famous flying ace, decided to pursue his dream of building a motorcar. A highly successful race driver before the war, Rickenbacker had developed certain ideas for the design of an automobile that would live up to his name. It had to have a high-speed motor; it had to have a low center of gravity for safety and comfort; and, as a fallout from his racing days, it had to have four-wheel brakes. This was all well and good, but a famous name and a set of lofty ideas neither contributed a cent toward the outfitting of a car factory nor filled it with manpower. In other words, Rickenbacker needed someone to put up the money if his dream were to come true.

In 1919, Rickenbacker contacted his friend, Harry Cunningham, and discussed his ambitions with him. Cunningham was a buddy from Rickenbacker's racing days. At one time, Cunningham had worked for Alexander Winton, helped Henry Ford build the famous racer "999," and then became a Ford Motor Company branch manager. In 1908, he transferred his allegiance from the Ford Motor Company to EMF, followed Flanders to Maxwell Motor Company, and had just left the latter company when Flanders retired.

Cunningham took Rickenbacker's dream to Everitt, Metzger, and Flanders to sound them out about starting another new car company. Metzger declined the invitation. He already was vice president of the Columbia Motor Car Company and was involved in so many private and civic groups that he had no more room on his plate for other ventures.

Everitt Enters the Rickenbacker Picture

Everitt definitely became interested and brought Flanders into the picture as a potential member of the board of directors if and when a company was incorporated.

These two could not keep themselves out of any automotive action. In 1919 alone, Everitt and Flanders had formed the L.D. Rockwell Company, a sales agency representing seven different parts manufacturers. Everitt was its president, and both were on the board. This was shortly after Flanders had left the Maxwell Motor Company to live a life of ease, as he had told others. Earlier in the year, Everitt had acquired the rights to build the Elbertz differential, an early form of traction control, and had begun production of it in the B.F. Everitt Company machine shop, another Everitt venture.[10-30]

Rickenbacker and his new partners decided that if they were to build a new high-performance car, it should have a six-cylinder engine but aim for the middle price field—between $1,500 and $2,000—to avoid competition with Ford.

According to Rickenbacker in his autobiography,[10-31]

> ...work started on the prototype early in 1920. The first two cars were built practically by hand. Although I kept in close touch, it was Cunningham, not I, who looked after the details. I was not even in Detroit.

Birth of the Rickenbacker Motor Company

The Rickenbacker Motor Company was incorporated in June 1921 with a capitalization of $5 million. Everitt was made president and general manager. Per Rickenbacker's autobiography,

> I realized that my greatest contribution would be in sales. I was therefore not the president of the Rickenbacker Motor Company, but the vice president and director of sales. Everitt was president.

Other officials of the new company and board members were Harry Cunningham, secretary and treasurer; Roy Hood, purchasing manager, a role he had played for Flanders at EMF, Studebaker, and Maxwell; and Carl Tichenor, formerly the production manager of the Pierce-Arrow Motor Car Company who had resigned from Pierce-Arrow to accept a similar position with Rickenbacker. The final board member was E.R. Evans, who had joined the group early and had helped Cunningham design the prototype. He became Rickenbacker's chief engineer. Evans was a consulting engineer who once had been associated with the Canadian branch of the Metzger Motor Car Company.[10-32]

Other than Tichenor and Captain Eddie, each of the key players and board members had at one time or another been associates of Everitt, Metzger, and Flanders.

Outsiders looking into the company no doubt were quite aware of the past successes of the EMF crowd and expected that with men such as this in the fold, the Rickenbacker could not help but be a success. Unfortunately, timing often means everything, and timing definitely was not on Rickenbacker's side. The nation was in the throes of a deep recession. Matters had become so severe that major manufacturers such as Buick, Dodge, Maxwell, Reo, and Studebaker were forced to close down production for a time at the beginning of 1921 because people simply quit buying automobiles. Several companies failed. Ford almost was one of them. Henry Ford had gone deeply into debt to finance the Rouge Plant and managed to pay his bills only by continuing to build cars and forcing his dealers to buy them. No doubt this practice caused some dealers to go bankrupt, but the Ford Motor Company was saved.

Regardless of the social situation, everything initially seemed to go well for the Rickenbacker Motor Company. By the end of summer, plans were made to have the car built in one of the Everitt Brothers Company plants, which was being vacated for this purpose. However, the enthusiastic response that the announcement of the new car received (despite its sketchy description) lulled management into believing that it could sell upward of 50,000 cars per year, which called for building 200 cars each day. This was twice the projected number that could be assembled within the working space of the Everitt plant. Everitt looked around town and in September purchased the Michigan Avenue factory that housed Disteel Wheel, a division of the Pressed Steel Company.[10-33]

More encouragement came from the reception given to the sale of Rickenbacker company stock. To quote Eddie Rickenbacker in his autobiography,

> In the fall of 1921, we sold about $5 million worth of stock to some 13,000 stockholders. My three partners and I retained about 25 percent of the total.

Assuming this was true, the trust in the Rickenbacker name must have been enormous for people to make an average investment of approximately $385 each in an unknown company, especially with the country just beginning to shake off the effects of a bad recession.

Perhaps it was the combined magic of the Rickenbacker name, the Everitt name, and now the Flanders name (because the directors of the Rickenbacker Motor Company had elected Flanders as board chairman) that drew investors.

Flanders was quite optimistic about the future. He looked forward to a large boost in the demand for automobiles come the spring of 1922 when the Rickenbacker car would be introduced to the market. Flanders said,[10-34]

> The real cause of lack of orders has been curtailed buying power of the farmer. As rapidly as that is restored, just so soon will conditions pick up, for the farmer, after all, is the largest of our users of motor cars and everything else.

As usual, Flanders' assessment was correct—not necessarily about the farmers, but about business "picking up" again the following year. Things were not as good as they had been in 1920 but were far better than 1921.

Introduction of the Rickenbacker Cars

The first Rickenbacker automobiles were introduced during the January 1922 New York Auto Show. As Captain Eddie himself put it[10-35]:

> The three displayed [at the New York Auto Show] represented not only the entire Rickenbacker line but also the entire output of the factory at that date. They were a touring car, which sold for $1,484; a coupe, for $1,885, and a sedan for $1,995. All three were the most handsome automobiles of that day...The new features—high-speed engine and low-slung body, among others—made sense to the car-buying public. People obviously believed our slogan, "A Car Worthy of Its Name," for the orders began pouring in.

Providentially, at the Rickenbacker exhibit, Captain Eddie met his future wife, Adelaide.

Although the new Rickenbacker did not have the four-wheel brakes that Captain Eddie wanted (his partners did not believe the public was ready for them yet), it did have everything else. Its 218 cubic inch six-cylinder engine developed 58 horsepower at 2800 rpm, and the company guaranteed that a Rickenbacker could travel at 60 miles per hour.

Unique to the Rickenbacker engine was the presence of a flywheel at either end of the crankshaft. This innovation was credited to Captain Eddie, who claimed that he had inspected the engine from a highly maneuverable enemy plane that he had shot down during the war and found that it had had a second flywheel at the rear end of the crankshaft. For what it was worth, the added flywheel did seem to add up to a smoother-running automobile, and the company touted it as a major selling point. (A man named George Lomb already had patented it.)

The frame rails of the car were wide and quite deep at eight inches. Semi-elliptic front springs of 36 inches, combined with unusually long rear springs of 57 inches, added to the excellent ride of the vehicle.[10-36]

At the end of July, Everitt announced that the 2,500th Rickenbacker car had rolled off the assembly line. He also added that the company would have to triple that amount to fill its orders for the remainder of the year.[10-37] The company did not achieve that goal, as Everitt reported to the shareholders at their first annual meeting in January 1923. However, the company did produce (and we surmise, sell) 5,000 cars, which brought a tidy profit, enough to declare a five-percent dividend.[10-38]

A new Model B was introduced in late December 1922. Other than a new clutch, it was essentially the same as its predecessor. One item of note was the replacement of the touring car with a phaeton, which had a top that was permanently bolted into place; that is, it could not be lowered. Everitt decided that because most drivers never put down their soft-tops, there was no need to give them the drop mechanism.[10-39] The Model B was not the only news emanating from Rickenbacker at the end of the year.

The Return of the Master Wordsmith, LeRoy Pelletier

In time to jump-start the sales publicity for the new Rickenbacker car in late December was the golden-penned LeRoy Pelletier, who rejoined his old partners, Everitt and Flanders, as director of sales promotion and publicity.

Pelletier had spent the preceding six years trying to make it on his own. In July 1913, he had purchased the machinery and stock of the electric car that had been built by the defunct Flanders Manufacturing Company. With these in hand, Pelletier formed the Tiffany Electric Car Company, then hired Flanders Manufacturing Company ex-general manager Don McCord to produce the car.[10-40] There were two models: the Deluxe selling for $2,750, and the Mignon selling for $2,500. When the public did not respond as he had hoped, Pelletier tried a different tact, announcing an open roadster called the Bijou to sell for only $750, a price too ridiculously low for an electric to make a profit. In March 1914, after asking Walter Flanders' permission, Pelletier changed the name of the company back to Flanders Electric, hoping to achieve some success by using Walter Flanders' name.[10-41] Try as he may, Pelletier could not make his magic with words in advertising and marketing translate into a boom in sales. By autumn of that year, the company went into receivership, and Pelletier departed to form his own ad agency.

Everitt's Body Business Is in the News Again

In May 1923, one of the larger items of automotive news was that the Everitt Brothers Company had merged with the year-old automobile body firm, the Trippensee Manufacturing Company. The new corporation was titled the Trippensee Closed Body Corporation. Frank Trippensee became president, Roland Everitt vice president, and Gordon Everitt secretary/treasurer. Prior to the merger, Barney Everitt had sold all of his stock in the Everitt company to his brothers to focus his energies on running the Rickenbacker concern.[10-42] What occasioned the merger was the decision by Everitt for Rickenbacker Motor Company to produce nothing but closed models in its future, and a ready source for these bodies was required. Open models would be sold only by special order.

Rickenbacker's Wish for Four-Wheel Brakes Becomes Reality

During late June 1923, Captain Eddie finally got the braking system he originally had in mind for his car. Everitt announced that beginning in July, half of the cars produced by the Rickenbacker Motor Company would carry four-wheel brakes. These would sell for $150 more than their two-wheel brake siblings.

Although the marketing of four-wheel brakes by Rickenbacker seemed to come out of the blue, their development had not. Ever since Rickenbacker had proposed them in 1919, developmental work had continued on them within the organization. As eventually offered, they were a mechanical type, with braking power applied through four internal expanding, cam-operated brake shoes.[10-43]

What soon angered Captain Eddie was the ugly campaign that followed from his competitors within the industry. In particular, a one-page ad from Studebaker claimed that four-wheel brakes were extremely dangerous. Rickenbacker later wrote[10-44]:

> In every community, all the other dealers and salesmen ganged up on the Rickenbacker car and its four-wheel brakes. Some said that the car would turn over on a curve when the brakes were applied. Others claimed that all four wheels would skid, rather than grip. Some said that the four-wheel brakes would stop the car too quickly, throwing the occupants up against the dashboard and injuring them.

Captain Eddie felt that the entire industry had joined hands to discredit Rickenbacker Motor Company through adverse publicity; however, his words were written several years later in hindsight. They were the words of a man who could not quite fathom what had gone wrong with his ambition to succeed in the auto industry. Actually, within days of the announcement of the availability of Rickenbacker cars with four-wheel brakes, approximately 1,500 orders for them were received, enough to propel the company from 83rd to 19th place within the industry. In the midst of this flurry of business, on July 6 the company celebrated the production of its 10,000th vehicle.

At the end of the year, Everitt grandly announced that because so many people had purchased Rickenbacker shares of stock, he decided to discontinue their sales and take approximately one million shares off the market because the company did not need the extra money.

Business was beginning to hum along nicely as 1924 hovered into view. Pelletier had colorful, bright-sounding ads featured throughout the media. During March,

Rickenbacker set an internal record of building and shipping 92 cars in one day. However, by the end of the year, business had tapered off and the final build tally for the year was only 7,187 units produced for sale. Nevertheless, Everitt announced a net profit of more than one-quarter million dollars. After factoring in taxes, however, the Rickenbacker books displayed a loss of $147,763.[10-45] Soon Everitt could be found searching for a likely partner for a merger, but he failed to locate viable candidates.

In June 1924, Everitt announced the models for the 1925 coming year. New to the lineup were the Vertical 8 Super-fine models. Each carried an L-head eight-cylinder engine that displaced 268 cubic inches and developed 70 horsepower at 3000 rpm. It had an unusual oil-tight compartment enclosing its camshaft to bathe the camshaft in oil at all times. Four models were available, the lowest price being the Sport Phaeton selling for $2,195, and the highest being the five-passenger sedan selling for $2,795.

The Downward Slide of Everitt and Rickenbacker Begins

What may have hurt Rickenbacker more than anything was another severe recession that blanketed the industry in 1925. It probably did not help when Everitt began the year by arranging a merger not, as we would expect, with another car company, but with the Trippensee Closed Body Corporation, ostensibly to guarantee the shipment of closed bodies which were rapidly capturing the public's imagination and purse strings. By mid-year, with things progressively turning sour, Everitt attempted to give sales a boost by dropping the price of the eight-cylinder models by $200 at the low end and $600 at the high end. It was a bad decision, made without consulting the dealer body, which had on hand a high inventory of these models that had been purchased from the company at the old prices. A number of dealers could not sustain the loss and failed. Despite the recession, Rickenbacker production reached its highest numbers ever—9,214 cars for 1925.

The year 1926 would be another matter. Captain Eddie later recollected,[10-46]

> Our dealers could not sell the car, and they were going broke. It was up to us to shore them up. I put all the money I had into the Rickenbacker Motor Company, and when that was

gone I went out and borrowed more. I borrowed from banks and suppliers. One single bank loan was $50,000. My total personal indebtedness came to $250,000.

It took several years, but Rickenbacker eventually repaid the entire amount.

In September 1926, Everitt suspended the sale of Rickenbacker stock which he had reinstituted in March 1924. He called the action a mere formality to reassess things. In June, Everitt lowered prices throughout the Rickenbacker line, again without consulting the dealer body, with the usual acrimonious results. The directors began to argue among themselves about how best to increase business and profits. This led to Captain Eddie tendering his resignation as vice president and head of sales in September 1926. He said,

> Here's where I get off.

Shortly afterward, both LeRoy Pelletier and Harry Cunningham also left the company.

In November, Barney Everitt convinced the Columbia Axle Company to take Rickenbacker Motor Company to court over a trivial bill. The company still was in good shape with seven dollars in assets for every one dollar in liabilities. Everitt's gambit was to have his company placed in a "friendly" receivership, with him (and the Security Trust Company) designated as one of the receivers so that it could be totally reorganized. As reported in *Automobile Topics,* the action was:[10-47]

> ...instituted as a means of straightening out some kinks in the readjustment program which has been under consideration for some time. Operation of the business will not be affected...and the plan for introducing new models, already announced to the dealers will go forward as projected.

Shortly thereafter, Everitt was reelected as president.

New models indeed were announced, but the business was affected—probably more by the resignation of Rickenbacker than anything else. Captain

Eddie may not have been the best sales manager in the business, but he had prestige and a magnetic personality. With him gone, customers may have lost faith in a company that was his only in name.

The 1927 models were advertised as the "6-70" (six cylinders and 70 horsepower), the "8-80" (eight cylinders and 80 horsepower), and the "8-90" (eight cylinders and 90 horsepower). They were considered to be the most handsome of the Rickenbacker cars yet produced, but it was too late.

In mid-February 1927, Everitt, as company president, turned to the Security Trust Company and himself as the receivers for permission to sell the Rickenbacker properties. Everitt was said to have lost a million dollars on his part of the Rickenbacker venture.[10-48] Total production of Rickenbacker cars during the five-year life of the company has been estimated to be approximately 34,500 units. So much for grand hopes. To think that the Rickenbacker organizers had turned down the purchase of the Everitt Brothers Company plant in 1921 simply because it did not have enough space in which to build more than 20,000 cars per year!

The Everitt Beat Goes On

It was not until July or early August of 1928 that the Rickenbacker plant itself finally was sold. Its purchaser? Everitt's old friend, Walter O. Briggs, who had joined Everitt and others in backing a new venture. This time it was to manufacture small, low-priced airplanes and flying boats. The firm was called the Verville Aircraft Company, adding another to the roll call of organizations that had appointed Everitt as their president.[10-49]

In mid-1929, Everitt's name came into the news once more. This time, it was an announcement that he had just been elected president of the Aerocar Company of Detroit. Other stockholders were Roy Chapin, Howard Coffin, and Walter O. Briggs. They had put together a pool of $250,000 to start the new company. The Aerocar was, in reality, a camping trailer designed by Glenn Curtiss. Everitt planned to build it and three other models: one to perform as a small school bus, another as a small 12-passenger regular bus, and the fourth for commercial use.[10-50]

Verville went out of business in 1932 after building a few planes. The Aerocar Company disappeared without much of a trace, probably also a victim of the Great Depression. Entering the 1930s, Barney Everitt was 58 years old—too early for most men to retire, but probably not for a man who had already had as many business adventures as he.

Epilogue

Walter E. Flanders

At eight o'clock Saturday morning, June 16, 1923, Walter Flanders died in a hospital near Newport News, Virginia, from the complications of a tragic automobile accident. He was only 52 years old. The scene of the accident was a road outside his 1,500-acre estate on the James River near Williamsburg, which he had purchased in late 1919.

The accident had happened three weeks previously. A friend was driving the car in which Flanders was riding. When this friend attempted to pass another car in front, he lost control of the car. It skidded into a ditch and turned over, pinning Flanders beneath it. Both of Flanders' legs were fractured. After being taken to the hospital, the normally robust health of Walter Flanders gave way, kidney problems developed, and he did not rally.[E-1]

At the time of his death, Flanders was chairman of the Rickenbacker Motor Company but apparently had left its active management to Everitt and the other company officers. His real pursuit had been that of a gentleman farmer. Flanders raised livestock and poultry. In addition to the Jamestown property, he owned a vast estate of more than a thousand acres off Green Lake near Pontiac, Michigan, on which he employed at times 300 to 400 men.[E-2] The huge house that Flanders built there in 1914 sits on a knoll overlooking the grounds. It still exists, having served variously as country club and nursing home over the years since Flanders' death.

William E. "Bill" Metzger

Bill Metzger died of a heart attack at 3:00 A.M. on April 11, 1933, in the midst of the Great Depression. He was only 64 years old, but he had been having health problems for the preceding four years. During the last year of his life, Metzger's condition had become serious. His mind had weakened. There was speculation that the worsening of Metzger's condition had been brought about by his flying his airplane against his doctor's orders. Metzger's death was considered providential, in light of his illness.

At the time of his passing, Metzger was director of nine different airplane organizations, two of which were the Stout Service and the Curtiss Flying Service. He also owned stock in Northwest Airways, which he had helped organize.[E-3]

Byron F. "Barney" Everitt

Barney Everitt passed away in the Harper Hospital in Detroit on Saturday, October 5, 1940, at age 67. His health had been failing for more than a year. Notice of his death was overshadowed in the Detroit newspapers on the following day because of the excitement aroused by the Detroit Tigers playing Cincinnati at home in the fifth game of the World Series. The series was knotted at two games apiece, and the Tigers would go on to win the Sunday game but eventually lose the series.[E-4]

Everitt had left the automobile business after the Rickenbacker failure and devoted his time to real estate development. His last known positions were that of president of the Heinz Development Company, which he assumed in 1938, and the Sampson Rubber Products Company.

E. LeRoy Pelletier

LeRoy Pelletier spent his remaining years after Rickenbacker touting Detroit as the air center of the world. He also helped promote huge amusement centers such as Luna Park and Coney Island. Pelletier is considered the person who introduced the practice of adding "midways" to large amusement venues such as world fairs and other large expositions.

Pelletier died in the Henry Ford Hospital in Detroit of a heart disease on September 5, 1938. He was 70 years old.[E-5]

Endnotes

Chapter One: From Carriages to Car Bodies

1-1.   Leake, *History of Detroit*, The Lewis Publishing Co., Chicago, 1912, p. 901.
1-2.   Lamm, Michael, and Holls, Dave, *A Century of Automotive Style*, Lamm-Morada Publishing Company, Stockton, 1997.
1-3.   Leake, *History of Detroit*, The Lewis Publishing Co., Chicago, 1912, p. 901.
1-4.   May, George Smith, *A Most Unique Machine: The Michigan Origins of the American Automobile Industry*, Eerdmans, Grand Rapids, 1975, p. 119.
1-5.   Glasscock, C.B., *The Gasoline Age*, The Bobs-Merrill Co., New York, 1937, p. 177.
1-6.   *The Detroit Free Press*, Sunday, March 10, 1901.
1-7.   May, George Smith, *A Most Unique Machine: The Michigan Origins of the American Automobile Industry*, Eerdmans, Grand Rapids, 1975, p. 116.
1-8.   Nevins, Allan, *Ford: The Times, The Man, The Company*, Charles Scribner's Sons, New York, 1954, p. 229. Note: The Detroit City Directory gives a 77–79 Brush Street address for Everitt Automobile Trimmings and Painting in 1903, but a Fort Street address for 1904.
1-9.   Borth, Christy, addendum featuring a short biography of Everitt, National Automotive History Collection, Detroit Public Library, undated.
1-10. "Detroit, The Home of the Automobile Industry," *The Detroit Journal*, May 28, 1904.
1-11. Lewis, Eugene W., *Motor Memories*, Alvid Publishers, Detroit, 1947, p. 211. Also, in Lamm, Michael, and Holls, Dave, *A Century of Automotive Style*, Lamm-Morada Publishing Company, Stockton, 1997, p. 27, Briggs is said to enter the auto industry as an upholsterer for

Everitt in 1904, then be "running the place by 1905." These authors also state that Briggs and the Fishers lived in the same boarding house and speculate whether it was through their mutual discussions and friendship that both parties subsequently formed their own (highly successful) auto body businesses.

1-12. *The Detroit Journal,* May 28, 1904.
1-13. *The Motor World,* XVII, March 12, 1908.
1-14. May, George Smith, *A Most Unique Machine: The Michigan Origins of the American Automobile Industry,* Eerdmans, Grand Rapids, 1975, p. 305.
1-15. Davis, Donald F., *Conspicuous Production: Automobiles and Elites in Detroit, 1899–1933,* Temple University Press, Philadelphia, 1988, p. 50.
1-16. *The National Cyclopaedia of American Biography,* Vol. VXXII, James T. White & Co., 1932, p. 222.
1-17. *The National Cyclopaedia of American Biography,* Vol. XVIII, James T. White & Co., 1932, p. 81.
1-18. Annual report of the Wayne Automobile Company, as of November 10, 1904, State Archives, Michigan Department of State.
1-19. Kimes, Beverly, and Clark, Henry Austin, Jr., *The Standard Catalogue of American Cars 1805–1942,* Krause Publications, Iola, Wisconsin, 1989, p. 1474.
1-20. May, George Smith, *A Most Unique Machine: The Michigan Origins of the American Automobile Industry,* Eerdmans, Grand Rapids, 1975, p. 306.
1-21. Annual report of the Wayne Automobile Company, as of December 29, 1906, State Archives, Michigan Department of State.
1-22. Kimes, Beverly, and Clark, Henry Austin, Jr., *The Standard Catalogue of American Cars 1805–1942,* Krause Publications, Iola, Wisconsin, 1989, p. 1474.

Chapter Two: The Making of Cadillac and Other Daring Deals

2-1. Burton, Clarence, and Burton, M. Agnes, *History of Wayne County,* S.J. Clarke Publishing Company, Detroit, 1930, p. 157.

2-2. Stark, George W., "On a Bicycle Built for Two," *D.A.C. News,* Vol. 14, July 1929.
2-3. *Illustrated Detroit,* August 1891, p. 156.
2-4. *Detroit Saturday Night,* April 22, 1928, p. 5.
2-5. Letter to Mr. David Beecroft, October 22, 1924, in King Collection, Beecroft Papers, National Automotive History Collection, Detroit Public Library.
2-6. Letter to Mr. David Beecroft, July 7, 1915, in King Collection, Beecroft Papers, National Automotive History Collection, Detroit Public Library.
2-7. Letter to Mr. David Beecroft, October 22, 1924, in King Collection, Beecroft Papers, National Automotive History Collection, Detroit Public Library.
2-8. Letter to Mr. David Beecroft, July 7, 1915, in King Collection, Beecroft Papers, National Automotive History Collection, Detroit Public Library.
2-9. Kimes, Beverly, and Clark, Henry Austin, Jr., *The Standard Catalogue of American Cars 1805–1942,* Krause Publications, Iola, Wisconsin, 1989, p. 1472.
2-10. Epstein, Ralph C., *The Automobile Industry,* A.W. Shaw Company, Chicago and New York, 1928, p. 96.
2-11. Letter to Mr. David Beecroft, December 7, 1915, in King Collection, Beecroft Papers, National Automotive History Collection, Detroit Public Library.
2-12. "It's Showtime," *D.A.C. News,* January 1987, p.20.
2-13. Nevins, Allan, *Ford: The Times, The Man, The Company,* Charles Scribner's Sons, New York, 1954, p. 204; *The Detroit News,* October 11, 1901; *The Detroit Free Press,* October 11, 1901.
2-14. "Retail Trade and Garages," *The Automobile,* December 12, 1903; and "Detroit, The Home of the Automobile Industry," *The Detroit Journal,* April 28, 1904, p. 8.
2-15. Letter to Mr. David Beecroft, July 7, 1915, in King Collection, Beecroft Papers, National Automotive History Collection, Detroit Public Library.
2-16. *Cycle and Automobile Trade Journal,* December 1, 1902, p. 66.
2-17. Letter to Mr. David Beecroft, July 7, 1915, in King Collection, Beecroft Papers, National Automotive History Collection, Detroit Public Library.

2-18. Kimes, Beverly, and Clark, Henry Austin, Jr., *The Standard Catalogue of American Cars 1805–1942*, Krause Publications, Iola, Wisconsin, 1989, p. 1525.
2-19. May, George Smith, *A Most Unique Machine: The Michigan Origins of the American Automobile Industry*, Eerdmans, Grand Rapids, 1975, p. 244.
2-20. *Ibid.*
2-21. Letter to Mr. David Beecroft, October 22, 1924, in King Collection, Beecroft Papers, National Automotive History Collection, Detroit Public Library.
2-22. Kimes, Beverly, and Clark, Henry Austin, Jr., *The Standard Catalogue of American Cars 1805–1942*, Krause Publications, Iola, Wisconsin, 1989, p. 190.
2-23. Crabb, Richard, *Birth of a Giant*, Chilton Book Co., Philadelphia, 1969, p. 102.
2-24. May, George Smith, *A Most Unique Machine: The Michigan Origins of the American Automobile Industry*, Eerdmans, Grand Rapids, 1975, p. 253.
2-25. *The Detroit Journal*, April 28, 1904.
2-26. May, George Smith, *A Most Unique Machine: The Michigan Origins of the American Automobile Industry*, Eerdmans, Grand Rapids, 1975, p. 255.
2-27. Davis, Donald F., *Conspicuous Production: Automobiles and Elites in Detroit, 1899–1933 (Technology and Urban Growth)*, Temple University Press, Philadelphia, 1988, p. 64.
2-28. Letter from William Metzger to Henry Joy, December 26, 1905, National Automotive History Collection, Detroit Public Library.
2-29. Thomas, L.M., "Following the Mystery Tide: The Life and Cars of Charles B. King," *Automotive Quarterly*, Vol. 30, No. 3, 1992.

Chapter Three: The Merry Master of Mass Production

3-1. Sorenson, Charles E., *My Forty Years with Ford*, W.W. Norton & Company, Inc., New York, 1956, p. 84.
3-2. *Ibid.*
3-3. MacManus, Theodore, *Men, Money, and Motors*, Harper & Bros., New York, 1929, p. 138.

3-4. Nevins, Allan, *Ford: The Times, The Man, The Company,* Charles Scribner's Sons, New York, 1954, p. 334.
3-5. *The Detroit News,* June 17, 1923.
3-6. Glasscock, C.B., *The Gasoline Age,* The Bobs-Merrill Co., New York, 1937, p. 118.
3-7. Sorenson, Charles E., *My Forty Years with Ford,* W.W. Norton & Company, Inc., New York, 1956, p. 45.
3-8. Doolittle, James R., ed., *The Romance of the Automobile Industry,* The Klebold Press, New York, 1916, p. 152.
3-9. Glasscock, C.B., *The Gasoline Age,* The Bobs-Merrill Co., New York, 1937, p. 188.
3-10. *Ibid.*
3-11. Nevins, Allan, *Ford: The Times, The Man, The Company,* Charles Scribner's Sons, New York, 1954, p. 336.
3-12. Sorenson, Charles E., *My Forty Years with Ford,* W.W. Norton & Company, Inc., New York, 1956, p. 96.
3-13. Nevins, Allan, *Ford: The Times, The Man, The Company,* Charles Scribner's Sons, New York, 1954, p. 364.
3-14. Sorenson, Charles E., *My Forty Years with Ford,* W.W. Norton & Company, Inc., New York, 1956, p. 88.
3-15. Simonds, William, *Henry Ford,* The Bobs-Merrill Co., New York, 1943, p. 107.
3-16. "Flanders Acquires Interest in Wayne," *The Motor World,* March 12, 1908.
3-17. Davis, Donald F., *Conspicuous Production: Automobiles and Elites in Detroit, 1899–1933 (Technology and Urban Growth),* Temple University Press, Philadelphia, 1988, p. 68.
3-18. Flanders' letter to the EMF Board of Directors, April 12, 1909, in the Minutes of the Everitt-Metzger-Flanders Company Directors Meeting of April 29, 1909, Studebaker Archives, Studebaker National Museum.
3-19. Davis, Donald F., *Conspicuous Production: Automobiles and Elites in Detroit, 1899–1933 (Technology and Urban Growth),* Temple University Press, Philadelphia, 1988, p. 69.

Chapter Four: EMF Bursts onto the Automotive Scene

4-1. "Wm. E. Metzger's Cat is out of the Bag at Last," *The Automobile,* June 4, 1908, p. 798.
4-2. "Michigan Men in a Big Deal," *The Motor World,* Volume XVIII, No. 10, June 4, 1908.
4-3. Kimes, Beverly, and Clark, Henry Austin, Jr., *The Standard Catalogue of American Cars 1805–1942,* Krause Publications, Iola, Wisconsin, 1989, p. 764.
4-4. *The Automobile,* June 4, 1908, p. 798.
4-5. "Big Merger Effected," *The Horseless Age,* June 3, 1908, p. 671.
4-6. "Metzger Motor Car Company," *Cycle and Automobile Trade Journal,* December 1909, p. 136.
4-7. Kimes, Beverly, and Clark, Henry Austin, Jr., *The Standard Catalogue of American Cars 1805–1942,* Krause Publications, Iola, Wisconsin, 1989, p. 764.
4-8. "The First Word About the New E.M.F. 30," *Motor Age,* July 30, 1908, p. 22.
4-9. *Ibid.*; also, "Details of the E-M-F '30,' " *The Automobile,* July 30, 1908, p. 161.
4-10. Glover, John G., and Cornell, William R., *The Development of American Industries,* Prentice-Hall, Inc., New York, 1951, p. 811; also, Doolittle, James R., ed., *The Romance of the Automobile Industry,* The Klebold Press, New York, 1916, p. 154.
4-11. *Ibid.*; Doolittle, James R., ed., *The Romance of the Automobile Industry,* The Klebold Press, New York, 1916, p. 155.
4-12. *Ibid,* p. 156.
4-13. Minutes of the First Meeting of the Stockholders of the Everitt-Metzger-Flanders Company, August 4, 1908, followed by the Meeting of the Board of Directors of the Everitt-Metzger-Flanders Company held August 4, 1908, upon the adjournment of the Stockholders Meeting, Studebaker Archives, Studebaker National Museum.
4-14. Flanders' letter to the EMF Board of Directors, April 12, 1909, in the Minutes of the Everitt-Metzger-Flanders Company Directors Meeting of April 29, 1909, Studebaker Archives, Studebaker National Museum.
4-15. Minutes of the Special Directors Meeting, Studebaker Brothers Manufacturing Company, July 10, 1908, Studebaker Archives, Studebaker National Museum.

4-16. Critchlow, Donald T., *Studebaker*, Indiana University Press, Bloomington, 1996, p. 51.
4-17. *Ibid*, p. 52.
4-18. *Ibid*, p. 55.
4-19. "Studebaker and E.-M.-F. Sales Combination, " *The Automobile*, August 13, 1908, p. 243; also, "E.-M.-F.—Sales Alliance," *The Horseless Age*, August 12, 1908, p. 211.
4-20. Flanders' letter to Metzger, Everitt, and Palms, dated July 29, 1908, attached to the Minutes of the Everitt-Metzger-Flanders Company Directors Meeting, April 29, 1909, Studebaker Archives, Studebaker National Museum.
4-21. *Ibid*.
4-22. Flanders' letter of August 28, 1908, Exhibit A of the Minutes of the Everitt-Metzger-Flanders Company Directors Meeting, April 29, 1909, Studebaker Archives, Studebaker National Museum.
4-23. Minutes of the Everitt-Metzger-Flanders Company Directors Meeting, September 15, 1908, Studebaker Archives, Studebaker National Museum.
4-24. Minutes of the Special Meeting of the Stockholders of the Everitt-Metzger-Flanders Company, September 17, 1908, Studebaker Archives, Studebaker National Museum.
4-25. Flanders' letter to the EMF Board of Directors, April 12, 1909, in the Minutes of the Everitt-Metzger-Flanders Company Directors Meeting of April 29, 1909, Studebaker Archives, Studebaker National Museum.
4-26. *Ibid*.
4-27. "Northern Merged into E-M-F Company," *The Automobile*, October 8, 1908, p. 517.
4-28. Yanik, Anthony J., "Alfred Owen Dunk: Owner of Almost 800 Automobile Companies," *Chronicle*, Vol. 26, No. 3, 1991, Historical Society of Michigan.
4-29. *The Automobile*, June 14, 1909; also, *Motor Age*, June 14, 1909, p. 23.
4-30. *Motor Age*, February 11, 1909, p. 36.

Chapter Five: EMF Loses Its "E" and "M"

5-1. Everitt's letter to the EMF Board of Directors, March 4, 1909, in the Minutes of the Special Everitt-Metzger-Flanders Company Board Meeting of March 8, 1909, Studebaker Archives, Studebaker National Museum.

5-2. *Ibid.*

5-3. Minutes of the Special Adjourned Meeting of the Directors of the Everitt-Metzger-Flanders Company of March 8, 1909, Studebaker Archives, Studebaker National Museum.

5-4. Fish's letter to Flanders, dated April 8, 1909, Exhibit E in the Minutes of the Everitt-Metzger-Flanders Directors Meeting of April 29, 1909, Studebaker Archives, Studebaker National Museum.

5-5. *Ibid.*

5-6. "Pathfinding for the Glidden Tour," Catalog File 47-72, National Automotive History Collection, Detroit Public Library.

5-7. "Sixth Annual A.A.A. Reliability Tour," *The Horseless Age,* Vol. 24, No. 2, July 14, 1909.

5-8. Flanders' letter to the EMF Board of Directors, April 12, 1909, in the Minutes of the Everitt-Metzger-Flanders Company Directors Meeting of April 29, 1909, Studebaker Archives, Studebaker National Museum.

5-9. *Ibid.*

5-10. Minutes of the Special Meeting of the Directors of the Everitt-Metzger-Flanders Company, April 21, 1909, Studebaker Archives, Studebaker National Museum. It is doubtful if EMF had a low-priced car on the drawing board at this time, in light of events that would take place a few months hence.

Chapter Six: Flanders Expands EMF

6-1. Letter from Metzger and Everitt to the EMF Board of Directors, in the Minutes of the Everitt-Metzger-Flanders Company Directors Meeting of April 29, 1909, Studebaker Archives, Studebaker National Museum.

6-2. Offer of Studebaker Automobile Company to the Everitt-Metzger-Flanders Company, April 29, 1909, Exhibit A in the Minutes of the Everitt-Metzger-Flanders Company Board Meeting of April 29, 1909, Studebaker Archives, Studebaker National Museum.

6-3. Resolution to the EMF Board of Directors, in the Minutes of the Everitt-Metzger-Flanders Company Board Meeting of April 29, 1909, Studebaker Archives, Studebaker National Museum.

6-4. "Studebakers Buy Interest in E-M-F Company," *The Horseless Age,* May 5, 1909, p. 651.

6-5. Kimes, Beverly, and Clark, Henry Austin, Jr., *The Standard Catalogue of American Cars 1805–1942,* Krause Publications, Iola, Wisconsin, 1989, p. 409.
6-6. "Studebaker Acquires De Luxe," *The Motor World,* July 22, 1909.
6-7. Studebaker-Flanders "20" advertisement in *The Detroit Journal,* July 24, 1909.
6-8. Kimes, Beverly Rae, "E & M & F and LeRoy," *Automobile Quarterly,* Vol. 17, No. 4, Fourth Quarter, 1979.
6-9. "E-M-F Buys De Luxe Plant—Will Build Small Car," *The Automobile,* July 21, 1909, p. 159; also, "The E-M-F Company Buys Deluxe," *The Horseless Age,* Vol. 24, No. 3, July 21, 1909, p. 78.
6-10. "E-M-F Buys Two Plants," *The Detroit Journal,* July 28, 1909, p. 1.
6-11. *Ibid.*
6-12. Doolittle, James R., ed., *The Romance of the Automobile Industry,* The Klebold Press, New York, 1916, p. 156.
6-13. "The E-M-F Company," *Cycle and Automobile Trade Journal,* December 1909, p. 141.
6-14. Nevins, Allan, *Ford: The Times, The Man, The Company,* Charles Scribner's Sons, New York, 1954, p. 409.
6-15. "The E-M-F Company," *Cycle and Automobile Trade Journal,* December 1909.
6-16. *Ibid.*
6-17. Flanders' letter to the Studebaker Automobile Company of December 6, 1909, in the Minutes of the Special Directors Meeting of the Everitt-Metzger-Flanders Company on December 31, 1909, Studebaker Archives, Studebaker National Museum.
6-18. Flanders' telegram to the Studebaker Automobile Company, November 30, 1909, in the Minutes of the Special Directors Meeting of the Everitt-Metzger-Flanders Company on December 31, 1909, Studebaker Archives, Studebaker National Museum.
6-19. Flanders' letter to the Studebaker Automobile Company of December 6, 1909, in the Minutes of the Special Directors Meeting of the Everitt-Metzger-Flanders Company on December 31, 1909, Studebaker Archives, Studebaker National Museum.
6-20. *Ibid.*
6-21. Flanders' letter to the Studebaker Brothers Manufacturing Company of December 9, 1909, as inserted in the Minutes of the Special Directors Meeting of the Everitt-Metzger-Flanders Company on December 31, 1909, Studebaker Archives, Studebaker National Museum.

6-22. Ibid.
6-23. "E-M-F-Studebaker Break," *The Motor World*, December 16, 1909.
6-24. "Walter E. Flanders—Industrial Colossus," an eight-page advertisement that appeared in the 1911 *Munsey Magazine* Advertising Section. The ad copy does not mention who the "sales organization that formerly acted as its distributor" was, but that "sales organization" could have been a reference only to Studebaker.
6-25. Doolittle, James R., ed., *The Romance of the Automobile Industry*, The Klebold Press, New York, 1916, p. 157.

Chapter Seven: Crisis or Comedy?

7-1. "E-M-F and Studebaker Interests Are at Variance," *The Automobile*, December 16, 1909, p. 1061.
7-2. "Studebaker and E-M-F Difficulties Still Unsettled," *The Automobile*, December 23, 1909, p. 1107; and "Legal Coup by Studebakers," *Motor Age*, December 23, 1909, p. 43.
7-3. "Injunction Against E-M-F Company Denied," *The Horseless Age*, Vol. 24, No. 26, December 29, 1909, p. 802.
7-4. Judge Swan's response to the motion for dismissal filed by the Studebaker Automobile Company, written December 29, 1909, the Circuit Court of the United States for the Eastern District of Michigan, Southern Division; letter from Court to both parties.
7-5. Minutes of the Special Directors Meeting of the Everitt-Metzger-Flanders Company on December 31, 1909, Studebaker Archives, Studebaker National Museum.
7-6. "New Suits in Studebaker—E-M-F Legal War," *Motor Age*, January 6, 1910, p. 92.
7-7. *Ibid.*; also, "E-M-F—Studebaker Separation Completed," *The Horseless Age*, January 5, 1910.
7-8. "Studebaker—E-M-F Case Back to Detroit," *The Horseless Age*, Vol. 25, No. 3, January 19, 1910, p. 126.
7-9. "Studebaker Case Comes Up Again," *Motor Age*, January 13, 1910, p. 40.
7-10. "No Injunction Against E-M-F," *The Motor World*, February 10, 1910.
7-11. Yanik, Anthony J., "U.S. Motor: Ben Briscoe's Shattered Dream," *Automobile Quarterly*, Vol. 36, No. 2, February 1997.

7-12. Lewchick, Wayne, *On the Line; Essays in the History of Auto Work*, University of Illinois Press, Chicago, 1989, p. 20.
7-13. Minutes of the Directors Meeting of the Studebaker Brothers Manufacturing Company, March 5, 1910, Studebaker Archives, Studebaker National Museum.
7-14. Minutes of a Special Directors Meeting of the Everitt-Metzger-Flanders Company, March 8, 1910, Studebaker Archives, Studebaker National Museum.
7-15. *Ibid*.
7-16. "Studebaker Buys E-M-F," *The Motor World*, March 10, 1910.
7-17. "Big Auto Deal Means Extension of Local Plant," *The Detroit News*, March 10, 1910.
7-18. "Studebaker's Purchase of E-M-F, via J.P. Morgan, Startles Trade," *The Horseless Age*, Vol. 25, No. 11, March 16, 1910, p. 412.
7-19. "E-M-F Company Bought by Morgan Interests," *The Detroit Free Press*, March 9, 1910.
7-20. Critchlow, Donald, *The Life and Death of an American Corporation*, Indiana University Press, Bloomington, 1996, p. 62.
7-21. Minutes of the Regular Everitt-Metzger-Flanders Company Board Meeting of April 5, 1910, Studebaker Archives, Studebaker National Museum.
7-22. "Garford Parts from Studebaker," *The Motor World*, July 26, 1910.
7-23. Minutes of the Adjourned Directors' Meeting of the Everitt-Metzger-Flanders Company, April 6, 1910, Studebaker Archives, Studebaker National Museum.
7-24. "Studebaker Acquires Control," *The Motor World*, May 6, 1910.
7-25. "Motor Car News," *The Detroit Journal*, August 5, 1910.
7-26. Kimes, Beverly Rae, "E & M & F and LeRoy," *Automobile Quarterly*, Vol. 17, No. 4, Fourth Quarter, 1979.
7-27. Undated advertisement for the Flanders "20," National Automotive History Collection, Detroit Public Library.
7-28. *Ibid*.
7-29. Critchlow, Donald, *The Life and Death of an American Corporation*, Indiana University Press, Bloomington, 1996, p. 62.
7-30. "Studebaker with $45,000,000," *The Motor World*, February 2, 1911; also, "Studebaker Issue Snapped Up," *The Motor World*, March 2, 1911.

7-31. Minutes of the Board of Directors of the Studebaker Corporation, May 2, 1911, Studebaker Archives, Studebaker National Museum.
7-32. "Walter E. Flanders—Industrial Colossus," 1911 advertisement in *Munsey's Magazine*, from the D. Cameron Peck Collection, National Automotive History Collection, Detroit Public Library.
7-33. "Flanders in Touring Form," *The Motor World*, March 16, 1911.
7-34. Studebaker advertisement in *The Motor World*, June 22, 1911.
7-35. "New E.M.F. Model," *Cycle and Automobile Trade Journal*, June 1, 1911, p. 150.
7-36. EMF advertisement in *The Motor World*, September 28, 1911, pp. 46–47.
7-37. "Flanders Forms a New Company," *The Motor World*, May 12, 1910.
7-38. "E-M-F Men Go into Motorcycles," *The Motor World*, October 20, 1910.
7-39. "Incorporation of Flanders Company to Merge Five Michigan Plants," *The Horseless Age*, January 18, 1911, p. 181; and "News of Maker and Dealer in Many Fields," *The Automobile*, January 19, 1911, p. 242.
7-40. Letter from Walter Flanders to Frederick Fish, in the Minutes of the Directors Meeting of the Studebaker Corporation, May 2, 1911, Studebaker Archives, Studebaker National Museum.
7-41. "Flanders Made President of His Own Company," *The Horseless Age*, December 16, 1911.
7-42. "Flanders and Universal Merged," *The Horseless Age*, Vol. 29, No. 3, January 17, 1912.
7-43. "Flanders to Change Name," *Cycle and Automobile Trade Journal*, November 1911, p. 91.
7-44. "Fish Becomes Studebaker Head," *The Motor World*, December 14, 1911.

Chapter Eight: The Rebirth of Everitt and Metzger

8-1. "The Metzger Motor Car Company," *The Horseless Age*, September 22, 1909, p. 321; also, "Metzger Motor Car Co. Organized in Detroit," *Motor Age*, September 23, 1909, p. 15; also, "Metzger Motor Car Company," *Cycle and Automobile Trade Journal*, December 1909, p. 136.

8-2. "The New Everitt '30,'" *The Automobile*, September 30, 1909, p. 567; also, "The Metzger Motor Car Co.," *Cycle and Automobile Trade Journal*, November 1909, p. 82; also, "The Everitt Thirty," *The Horseless Age*, December 8, 1909, p. 658; also, "The Everitt Thirty," *Cycle and Automobile Trade Journal*, December 1909, p. 186; and "The 'Everitt 30' for 1910," *The Automobile*, January 20, 1910, p. 146.

8-3. Kimes, Beverly, and Clark, Henry Austin, Jr., *The Standard Catalogue of American Cars 1805–1942*, Krause Publications, Iola, Wisconsin, 1989.

8-4. "Metzger Absorbs Hewitt Entire," *The Motor World*, January 6, 1910.

8-5. Mroz, Albert, *The Illustrated Encyclopedia of American Trucks and Commercial Vehicles*, Krause Publications, Iola, Wisconsin, 1996, p. 194.

8-6. "Description and 1911 Specifications of the 'Everitt 30,'" *Cycle and Automobile Trade Journal*, June 1911, p. 142; and "Everitt 30 for 1911," *The Horseless Age*, February 1, 1911, p. 250.

8-7. "Hewitt and Metzger Part Company—New Million Dollar Corporation," *The Horseless Age*, December 27, 1911.

8-8. "Metzger and Hewitt Unmerged," *The Motor World*, December 21, 1911.

8-9. Mroz, Albert, *The Illustrated Encyclopedia of American Trucks and Commercial Vehicles*, Krause Publications, Iola, Wisconsin, 1996, p. 194.

8-10. "Evidence of Everitt Expansion," *The Motor World*, October 5, 1911.

8-11. Advertising booklet, 31 pages, "Story of the Everitt," National Automotive History Collection, Detroit Public Library.

Chapter Nine: Flanders Reunites with Everitt and Metzger

9-1. "Studebakers Start Their Moves," *The Motor World*, January 18, 1912.

9-2. "Flanders Admits His Resignation," *The Motor World*, February 8, 1912.

9-3. "Flanders Pot Keeps Boiling," *The Motor World*, February 1, 1912.

9-4. "Late Details of Everitt-Metzger-Flanders Deal," *The Automobile*, May 16, 1912.

9-5. "Studebaker in Annual Session," *The Motor World*, April 4, 1912.

9-6. "Metzger Affairs Looming Large," *The Motor World*, May 9, 1912.

9-7. "Late Details of Everitt-Metzger-Flanders Deal," *The Automobile*, May 16, 1912; also, "Flanders and Metzger Companies Merged," *The Horseless Age*, May 15, 1912; also, "Flanders in the Metzger Deal," *The Motor World*, May 16, 1912.

9-8. Letter from Paul Smith dated May 4, 1912, in the Minutes of the Directors Meeting of the Everitt Motor Car Company, May 22, 1912, Chrysler Archives, DaimlerChrysler AG.

9-9. "Good-bye to E-M-F and Flanders," *The Motor World*, August 8, 1912.

9-10. "Everitt Cars Renamed Flanders," *The Motor World*, September 12, 1912.

9-11. Yanik, Anthony J., "U.S. Motor," *Automobile Quarterly*, Vol. 36, No. 2, February 1997.

9-12. "United States Motor Company Awaits Action of Its Creditors," *The Automobile*, September 19, 1912.

9-13. Yanik, Anthony J., "U.S. Motor," *Automobile Quarterly*, Vol. 36, No. 2, February 1997.

9-14. "Opposition to U.S. Motor Reorganization Develops," *The Motor World*, October 24, 1912.

9-15. "To Sell United States Motor Properties on January 8," *The Horseless Age*, November 20, 1912; and "Walter E. Flanders Becomes President of the Reorganized United States Motor Company," *Automobile Trade Journal*, December 1, 1912.

9-16. "Exodus from U.S. Motor as New Regime Approaches," *The Motor World*, December 12, 1912.

9-17. "Smith and Pelletier Quit U.S. Motor Co.," *The Horseless Age*, January 1, 1913.

9-18. "Flanders and Chief Aids Reach Parting of Ways," *The Motor World*, January 2, 1913.

9-19. "Smith and Pelletier Quit U.S. Motor Co.," *The Horseless Age*, January 1, 1913.

9-20. "Only One Bidder for U.S. Motor Property," *The Horseless Age*, January 8, 1913.

9-21. "Maxwell Motor Co. to Succeed Standard," *The Horseless Age*, January 15, 1913; also, "Reorganized U.S. Motor Assumes Maxwell Name; Adds $6,000,000," *The Motor World*, January 16, 1913.

9-22. "U.S. Motor Reorganization Blocked by Indiana," *The Motor World*, January 30, 1913.
9-23. "The Maxwell Motor Co. Policy," *The Horseless Age*, February 3, 1913.
9-24. "Flanders, of Pontiac, Fails; Two Millions Gone to Waste," *The Motor World*, December 12, 1912.
9-25. "Dissension Places Flanders Manufacturing Co. in Receiver's Hands," *The Horseless Age*, December 11, 1912; also, "Receiver for Flanders Mfg. Co.," *The Automobile*, December 12, 1912.
9-26. "Flanders Companies' Pact Ordered Broken by the Court," *The Motor World*, December 26, 1912.
9-27. "Assets of Flanders Shrink More than Million Dollars," *The Motor World*, April 3, 1913; also, "To Sell Assets of Flanders Mfg. Co.," *The Horseless Age*, April 9, 1913.
9-28. "Flanders Property Sold," *The Horseless Age*, June 18, 1913.

Chapter Ten: "E" and "M" and "F" After 1913

10-1. "Detroit Athletic Club," *Automobile Quarterly*, Vol. 38, No. 1, July 1998.
10-2. Yanik, Anthony J., "Alfred Owen Dunk: Owner of Almost 800 Automobile Companies," *Chronicle*, Vol. 26, No. 3, 1991, Historical Society of Michigan.
10-3. Annual report to the state of Michigan, February 28, 1917, Columbia Motor Car Company.
10-4. Kimes, Beverly, and Clark, Henry Austin, Jr., *The Standard Catalogue of American Cars 1805–1942*, Krause Publications, Iola, Wisconsin, 1989.
10-5. Applequist, H.A. "The American Motorcar Industry of 1920," *Automobile Topics*, September 1956.
10-6. "Columbia Looks for Best Month in July," *Automobile Topics*, July 8, 1922.
10-7. Kimes, Beverly, and Clark, Henry Austin, Jr., *The Standard Catalogue of American Cars 1805–1942*, Krause Publications, Iola, Wisconsin, 1989.
10-8. "Liberty Motors Sale Confirmed by Court," *Automobile Topics*, September 29, 1923.

10-9. "Liberty Plant Sale May Be Held in June," *The Motor World*, May 16, 1923.
10-10. "Court's Permission to Sell Columbia Co. Will Be Sought," *Motor Age*, September 25, 1924; also, "Columbia Sale Halts Reorganization Hope," *The Motor World*, October 24, 1924.
10-11. "Two Old Timers Get Wills in Michigan," *Automobile Topics*, May 14, 1921.
10-12. Woodward, Jack, "Childe Harold Wills," *Special-Interest Autos*, August–October, 1977.
10-13. Mroz, Albert, *The Illustrated Encyclopedia of American Trucks and Commercial Vehicles*, Krause Publications, Iola, Wisconsin, 1996.
10-14. "A.A.A. Is Expected to Absorb the N.M.A." *Automobile Topics*, February 9, 1924.
10-15. Burton, Clarence, "William E. Metzger," *City of Detroit*, S.J. Clarke Publishing Company, Detroit, 1922; also, Burton, Clarence, "William E. Metzger," *History of Wayne County*, S.J. Clarke Publishing Company, Detroit, 1930; also, "W.E. Metzger Dies in Detroit Home," *Automobile Topics*, April 15, 1933; also, "William E. Metzger, Automotive Pioneer," *Automotive Industries*, April 15, 1933.
10-16. Glover, J.G., and Cornell, W.B., *The Development of American Industries*, Prentice-Hall, Inc., New York, 1951, p. 838.
10-17. "New Northville Aircraft Plant Backed by Financial Leaders," *The Northville Record*, June 7, 1929.
10-18. Kimes, Beverly, and Clark, Henry Austin, Jr., *The Standard Catalogue of American Cars 1805–1942*, Krause Publications, Iola, Wisconsin, 1989, p. 900.
10-19. "To Centralize Maxwell Plants," *The Horseless Age*, April 9, 1913.
10-20. "America at the Wheel," special edition by *Automotive News*, September 21, 1993; Tables of U.S. Car/Sales Production.
10-21. "McGuire Relinquishes Office in Maxwell Motor," *The Motor World*, June 12, 1913.
10-22. "Old Maxwell-Briscoe Tarrytown Plant Sold," *The Horseless Age*, July 8, 1914.
10-23. "$1,505,467 Surplus for Maxwell Motor Co.," *The Horseless Age*, October, 1914.
10-24. Maxwell advertisements for 1915.
10-25. Maxwell advertisement, summer 1916.
10-26. "Maxwell Low-Priced Marvel," *Motor West*, August 15, 1916.

10-27. "Flanders Moves On to Bigger Things," *Automobile Topics*, December 15, 1917.
10-28. "Two Detroit Body Companies Combine," *Automobile Topics*, May 19, 1923.
10-29. "Everitt Takes Over Springfield Body," *Automobile Topics*, August 11, 1917.
10-30. "B.F. Everitt Puts Up Differential," *Automobile Topics*, March 22, 1919.
10-31. Rickenbacker, Edward V., *RICKENBACKER*, Prentice-Hall, Inc., Englewood Cliffs, New Jersey, p. 144.
10-32. " 'E' and 'F' Again Unite to Back Rick," *Automobile Topics*, July 30, 1921.
10-33. Kimes, Beverly Rae, "The Rickenbacker," *Automobile Quarterly*, Vol. 13, No. 4, 1975.
10-34. "Flanders Looks to Big Spring Demand," *Automobile Topics*, November 12, 1921.
10-35. Rickenbacker, Edward V., *RICKENBACKER*, Prentice-Hall, Inc., Englewood Cliffs, New Jersey, p. 145.
10-36. Kimes, Beverly Rae, "The Rickenbacker," *Automobile Quarterly*, Vol. 13, No. 4, 1975.
10-37. "Celebrate 2,500th Rickenbacker Car," *The Motor World*, August 5, 1922.
10-38. "Rickenbacker Made 5,000 Cars in a Year," *The Motor World*, January 10, 1923.
10-39. "Rickenbacker Series B Cars for 1923," *The Motor World*, December 20, 1922.
10-40. "Flanders Electric About to Become 'Tiffanyized,'" *The Motor World*, July 21, 1913.
10-41. Kimes, Beverly, and Clark, Henry Austin, Jr., *The Standard Catalogue of American Cars 1805–1942*, Krause Publications, Iola, Wisconsin, 1989.
10-42. "New Trippensee Body Is Formed by Merger," *The Motor World*, May 30, 1923; also, "Trippensee Provides for Greater Output," *Automotive Industries*, June 7, 1923.
10-43. "Rickenbacker Adds Four Wheel Brake Model," *The Motor World*, June 27, 1923; also, *Automotive Industries*, June 28, 1923.
10-44. Rickenbacker, Edward V., *RICKENBACKER*, Prentice-Hall, Inc., Englewood Cliffs, New Jersey, p. 148.

10-45. Kimes, Beverly Rae, "The Rickenbacker," *Automobile Quarterly*, Vol. 13, No. 4, 1975.
10-46. Rickenbacker, Edward V., *RICKENBACKER*, Prentice-Hall, Inc., Englewood Cliffs, New Jersey, p. 148.
10-47. "Court Puts Everitt in Charge," *Automobile Topics*, November 6, 1936.
10-48. Borth, Christy, short monograph on Everitt, National Automotive History Collection, Detroit Public Library, undated.
10-49. "Everitt Heads Plane Company," *Automobile Topics*, August 11, 1928.
10-50. "Aerocar Company Plans Production," *Automobile Topics*, June 15, 1929.

Epilogue

E-1. "W.E. Flanders Dies of Hurts," *The Detroit Free Press*, June 17, 1923.
E-2. "Walter E. Flanders Dies in Virginia," *Automobile Topics*, June 23, 1923.
E-3. "Auto Pioneer Passes," *The Detroit News*, April 11, 1933; also, "W. E. Metzger Dies in Detroit Home," *Automotive Topics*, April 15, 1933; and C.H. Burton, *History of Wayne County*, S.J. Clarke Publishing Company, Chicago, 1930.
E-4. "Death Takes B.F. Everitt," *The Detroit News*, October 6, 1940.
E-5. "E & M & F...and LeRoy," *Automobile Quarterly*.

Index

Page numbers followed by *p* indicate a photo.

Aerocar Company of Detroit, 203, 204
Alden-Sampson Manufacturing Company, 116, 161
American Automobile Association (AAA), 182
American Electric Car Company, 176
American Motor Car Manufacturer's Association (AMCMA), 46
American Vehicle Company, 17
Anthony, M.B., 166
Association of Licensed Automobile Manufacturers (ALAM), 46, 147, 148
Aster, 15
Austin, L.A., 70
Auto Crank Shaft Company, 96
Auto Parts Manufacturing Company, 70, 174–175
Auto shows, 19, 72, 127

Baker electric car, 19
Barbour, George Jr., 23
Barbour, William
    Northern beginnings and, 21, 22, 23, 32
    officer in EMF, 47
    partnership with Flanders, 135, 137
Barthel, Oliver E., 27
Bennett, Albert, 47
Benz, 15, 16
Bi-mobile, 136
Bicycles
    Metzger's interest in, 12–13, 16–17
    popularity of, 12, 13
Black, Clarence, 31

Blomstrom, C.H., 92
Board of Fire Commissioners, Detroit, 182
Book, James
    background, 7–8
    EMF investment, 57
    officer in EMF, 47
    partnership with Flanders, 135, 137
    Wayne involvement, 6
Bowen, Lem, 31
Briggs, Walter O.
    airplane interests, 203
    background, 5–6
    Flanders Manufacturing and, 140
Briggs-Detroiter Company, 175
Briscoe, Ben
    Maxwell partnership, 23
    United States Motor and, 116, 161, 162
Briscoe Manufacturing Company, 161, 186
Brown & Sharp Manufacturing Company, 162
Brown, Scott, 156
Brownson, Robert M., 135, 137, 139
Brush, Alanson P., 27
Brush cars, 166
Brush Runabout Company, 116, 161
Buick, 37

Cadillac Aircraft Corporation, 184
Cadillac Automobile Company
    company background, 26–27
    first-year sales, 29
    merger with Leland, 30–31
    sales contract with Metzger, 27, 29–30
Cadillac Motor Car Company, 31
Campeau, Daniel, 19
C.H. Little Company, 5, 6
C.H. Wills & Company, 180
Chalfant, E.P., 46
Chalmers, Hugh, 174, 191
Chalmers Motor Corporation, 190–191
Chapin, Roy, 174

Index

Coffin, Howard, 184
Cole, R.E., 179
Columbia Axle Company, 202
Columbia Motor Car Company
    capitalization, 175
    drain of Liberty acquisition, 179–180
    electric car, 19
    product line and prices, 176–178, 177*p*
    purchase of Liberty Motor, 179
    success in early 1920s, 178
    suppliers, 176
    United States Motor and, 116, 161, 166
Columbia Six, 177*p*
Commercial Engineering Company of Detroit, 157
Conference on Uniform Traffic Laws, 183
Consolidated Manufacturing Company, 26
Cost of cars. *See* Prices of cars
Couzens, James, 42
C.P. Malcolm & Co., 22
C.R. Wilson Carriage Company, 2, 3
Cunningham, Harry, 140, 194, 202

Daimler, 15, 16
Dayton Motor Car Company, 116, 161
De Luxe Motor Car Company, 92–93, 175
Delafield, Frederick, 122
Detroit Athletic Club (DAC), 174
Detroit Auto Show, 19
Detroit Automobile Company, 21
Detroit Aviation Society, 183
Detroit Electric Car Company, 71
Detroit Motor Car Company, 175
Detroit Stove Works, 7, 22
Detroit Trust Company, 171, 172
Dodge brothers, 157
Duffield, Henry, 108
Dunk, Alfred Owen, 70–72, 71*p*, 174, 175
Duquesne Motor Car Company, 46
Duryeas, 16

227

Eames, Hayden, 90, 91
Edward G. Budd Manufacturing Company, 180
Electric cars
    by Baker and Columbia, 19
    Detroit Electric, 71
    by Flanders Manufacturing, 139, 170
    Studebaker products, 62
    by Tiffany, 199
EMF 30
    model improvements (1912), 131–132
    Pathfinder performance, 81–83
    photos, 51*p*, 53*p*, 55*p*, 73*p*, 82*p*, 104–105*p*, 126*p*, 132*p*, 134*p*
    plans to build, 48
    pricing plan (1910), 52, 65, 123
    production targets (1910), 97–98
    self-sufficient manufacturing goal, 53–54
    technical features, 50–52
EMF Company. *See* Everitt-Metzger-Flanders (EMF) Company
Evans, E.R., 195
Everitt 30
    features, 144–146
    photos, 145*p*, 149*p*, 154*p*
Everitt Brothers Company, 192–193
Everitt, Byron F. "Barney," 2*p*, 192*p*
    anger at Studebaker contract, 75–76
    body business, 192–193, 199
    corporate career, 1–3
    death, 206
    EMF role, 47
    EMF salary, 58
    EMF severance agreement, 89–90
    end of Flanders management role, 167–168
    Ford association, 5
    last business ventures, 203–204
    Metzger Motor founding, 143–144
    Metzger Motor stock, 148
    Olds contract, 3, 4
    Rickenbacker association, 194, 195, 196, 201
    sales arrangement change request, 77
    Studebaker buyout, 88
Everitt, Donna F. Shinnick, 3, 9
Everitt "Four-36," 153

Index

Everitt, Gordon, 199
Everitt-Metzger-Flanders (EMF) Company, 60*p*
    announcement of birth, 45
    capitalization, 47, 57
    company officers (1908), 47
    company status (1910), 125
    De Luxe purchase, 92–93
    EMF 30 production. *See* EMF 30
    Everitt and Metzger severance agreement, 89–90
    first mechanical assembly line, 117
    first-year production, 92
    Flanders 20. *See* Flanders 20
    Flanders' defense of progress, 84–85
    Flanders' push for high production, 59, 61
    Flanders' resignation, 159–160
    Ford plant purchase, 134
    founding shareholders, 56
    J.P. Morgan pullout crisis, 128
    J.P. Morgan purchase of shares, 117–118
    J.P. Morgan purchase reports, 118–120
    mass production innovations, 55
    Northern Motor purchase, 68–69, 70
    post-Everitt/Metzger reorganization, 90–92
    post-Studebaker sales contracts, 103
    production schedule (1910), 122–123
    production shortfall (1908), 72–73
    production targets (1910), 97–98
    replacement parts deal with Dunk, 70
    sales to dealers during court cases, 114
    Studebaker and
        absorption into Studebaker, 133
        benefits of arrangement, 66–67
        buyout of Everitt and Metzger, 88
        contract abridgement suit. *See* Studebaker vs. EMF
        contract proposal (for 1909–1910), 86–88
        Everitt's anger over deal, 77–78
        Flanders' argument for contract, 85–86
        Flanders' confidence in deal (1908), 65–66
        Flanders' later dissatisfaction with deal, 100–102
        lack of control over agents, 75–76, 77–78
        sales contract (1909), 61–62, 64–65, 79–80
        slowdown in receipts, 99–100

Everitt-Metzger-Flanders (EMF) Company *(continued)*
    Studebaker and *(continued)*
        Studebaker reorganization effects, 129
        unilateral ending of contract, 102
    supplier plant additions, 95–96
    Wayne memorandum of agreement details, 56–59
    Wayne Motor purchase, 67–68
    workforce loyalty, 98
Everitt Motor Car Company
    Everitt 30, 144–146, 145*p*, 149*p*, 154*p*
    evolution from Metzger Motor, 158–159
    name change to Flanders, 160–161
Everitt, Roland, 192, 199
Everitt "Six-48," 152

Federal Manufacturing Company, 62
Federal Motor Truck Company, 182
Fish, Federick
    business with Garford, 63–64
    dealings with Morgan, 117–118
    election to EMF board, 90
    reaction to EMF contract concerns, 79–80
    sales arrangement with EMF, 61
    suit against EMF. *See* Studebaker vs. EMF
Fisher Body Company, 4
Fisher, Fred and Charles, 3–4
Flanders 20
    endurance runs, 126–127
    features and prices (1911), 125–126, 130
    photos, 131*p*
    pricing plan (1910), 93
    see also Studebaker-Flanders 20
Flanders Manufacturing Company, 137*p*
    business problems (1912), 168–169
    capitalization and officers, 136–137
    end of business, 170–172
    establishment of, 136
    motorcycle and electric car projects, 139, 169–170
    stockholders, 140
    Studebaker contract, 138–139, 140

Flanders Motor Car Company
    incorporation from Everitt Motor, 160–161
    Pelletier and Smith resignation, 165
    purchase by Maxwell Motor, 167–168
    Standard and Maxwell connection, 166
Flanders, Walter
    background, 36–37
    business success, 121
    De Luxe purchase, 93
    death, 205
    described, 35–36, 36*p*
    dissatisfaction with Northern purchase, 68–69
    EMF and
        contract as president, 91
        faith in sales capability, 99
        manufacturing vision, 53–54
        officer position, 47
        production schedule (1910), 122–123
        production targets (1910), 97–98
        push for high production, 59, 61
        resignation, 159–160
        salary, 58
    Ford association
        departure from Ford, 41–43
        first business with, 37
        hiring by Ford, 38–39
        improvements to Model N sales, 39–40
    involvement in companies, 168
    jig job invention, 54
    manufacturing company establishment, 136
    mass production innovations, 55
    Maxwell Motor association
        beginnings of, 166
        financial success from, 190
        retirement, 191–192
    move to Everitt Motor, 160
    move to Wayne, 42
    parts companies purchases, 135–136
    production scheduling introduction, 40, 41
    replacement parts deal with Dunk, 70
    Rickenbacker Motor involvement, 197

Flanders, Walter *(continued)*
   Studebaker and
      agreement with new contract, 87
      confidence in deal, 65–66
      dissatisfaction with Studebaker (1909), 100–102
      employment contract, 138
      explanation of contract situation, 85–86
      reaction to Everitt letter, 78
      resignation from, 156–157
      response to sales situation, 84–85
      rumors of a departure, 140–141
      unilateral ending of contract, 102
   United States Motor contract, 163–164
   Wayne production recommendations, 43–44
Ford, Henry
   desire to increase production, 37–38
   failed companies, 20, 21
   racing interest, 5, 20, 26
Ford Motor Company
   association with Everitt, 5
   departure of Flanders, 41–43
   hiring of Flanders, 38–39
   Model N improvements, 39–40
   Piquette plant consolidation, 41
   Piquette sale to EMF, 134
"Fore door" look, 130, 132p, 149
Four-wheel brakes, 199–200
Fournier, Henry, 20

Garford, Arthur, 62–64, 66
General Motors, 128
Geneva steam car, 19
Glasscock, C.B., 3
Glidden Tour, 80–82, 83p, 97, 130
Goldman, Henry, 128, 129
Goldman, Sachs & Co., 128, 129
Grant & Wood Manufacturing Company, 135
Grant Automatic Machine Co., 135
Gray, John S., 38
Gray, William, 2

Index

Grey Motor Company, 161
Grosse Pointe race, 20
Gunderson, George B., 22, 23, 32
Gunderson, Victor, 32
Gunn, James, 140, 156

Harris Brothers & Co., 172
Haynes, 17
Heaslet, James, 94, 123
Henry Ford Company, 20
Hewitt, Edward R., 148
Hewitt Motor Company, 147–148, 150
Hood, Roy, 195
Hough, Charles M., 164, 165
House of Morgan. *See* J.P. Morgan & Company
Hower, F.B., 81
Huber & Metzger, 13, 14*p*
Huber, Stanley B., 13
Hudson, J.L., 11
Hurlburt Motor Truck Company, 180
Hurlburt, William, 180

Indiana Bicycle Company, 17
International Motor Company (IMC), 150

Johnson, Hugh, 2
Joy, Henry B., 174
J.P. Morgan & Company
    additions to Studebaker board, 122
    news reports of EMF purchase, 118–120
    pullout from EMF, 128
    purchase of EMF shares, 117–118
    speculations over dealings, 120
Judson, Ross, 184

Kellerman, Annette, 145*p*
Kelly, William
    background, 48–49
    EMF salary, 58
    partnership with Everitt and Metzger, 143–144
    severance agreement, 89–90
    Wayne shares, 8
Kidder, Peabody and Company, 180
King, Charles Brady, 22, 23, 25, 32
Kirchner, Otto, 108
Kirk Manufacturing Company, 25–26, 92
Kleinwort Sons & Co., 129
Kopmeier, Waldemar, 140

L.D. Rockwell Company, 194
League of American Wheelmen, 12
Lehman Brothers, 129
Leland & Faulconer Manufacturing Company, 28, 31
Leland, Henry M., 27, 30–31
Lemly Appraisal Company, 57
Lewis, Dai, 81
Liberty Motor Car Company, 179
Lomb, George, 198
Long, Ray, 175

Mack Trucks, 150
Maguire, William, 164, 166, 186
Malcolmson, Alexander T., 5
Mass production
    first mechanical assembly line, 117
    Ford's hiring of Flanders and, 38–39
    high-volume cost benefits, 131
    innovations from EMF, 30, 55
    jig job invention, 54
Maxwell, Jonathan
    design experience, 22
    Northern involvement, 23, 32
    partnership with Briscoe, 23, 25
    United States Motor exit, 164

Index

Maxwell Motor Company
    business success (1915–1917), 187, 189*p*, 190
    Chalmers lease agreement, 190–191
    Flanders Motor purchase, 167–168
    Flanders' retirement, 191–192
    Flanders' revival of, 185
    product line (1914), 185–186
    product line changes (early 1913), 166, 167
    sale of United plants, 186–187
    start as Standard Motor, 166
    touring car, 188*p*
Maxwell-Briscoe Motor Company, 25, 46, 116, 161
McCarroll, H.G, 184
McCord, Don, 199
Metzger Motor Car Company, 151*p*
    business rumors (1912), 158
    Everitt 30 features, 144–146
    evolution into Everitt Motor, 158–159
    factory complex, 153
    founding and capitalization, 143–144, 148
    Hewitt Motor merger, 147–148
    officers, 144
    product line (1911), 149, 149*p*, 152, 153
    production shortfall (1910), 146–147
    production targets (1911), 150
    six-cylinder model, 152
Metzger, William E. "Bill"
    auto dealership beginning, 17–18
    auto dealership expansion, 21
    auto interest beginnings, 13–16
    Auto Parts presidency, 174–175
    auto show initiation, 19
    background and bicycle career, 11–13, 12*p*, 14*p*, 16–17, 18*p*
    Cadillac association
        promotion schemes, 28–29
        sale of agency to, 31
        sales contract, 27, 29–30
        shares in, 31
    Columbia Motor involvement, 176
    commercial aviation interests, 183–184
    DAC founding, 174
    death, 205–206

Metzger, William E. "Bill" *(continued)*
    EMF investment, 57
    EMF salary, 58
    EMF severance agreement, 89–90
    end of management role in Flanders, 167–168
    Federal Motor directorship, 182
    Grosse Pointe race, 19–20
    Metzger Motor founding, 143–144
    Metzger Motor stock, 148
    Northern Manufacturing founding, 21, 22, 23
    Northern Motor shares, 32
    officer in EMF, 47
    Studebaker buyout, 88
    volunteer activities, 182–183
    Wills Sainte Claire association, 180, 182
    Yale car interest, 25–26
Michigan State Council of Motor Clubs, 182
Michigan Stove Company, 7, 22
Miller, John, 108, 109
Mitchell, W. Ledyard, 191
Mobile Company, 17
Model A, C, H, F (Wayne), 8, 9*p*
Model B, C, K, L (Northern), 33
Model N (Ford), 38, 39–40
Mohrhardt, J.S., 175
Monroe Body Company, 95, 96
Motorcycles by Flanders Manufacturing, 136, 169
Murphy, William, 26, 31

National Air Transport Company, 183
National Automobile Chamber of Commerce, 183
National Conference on Municipal Traffic Codes, 183
Northern Manufacturing Company
    company start, 21–22
    incorporation and capitalization, 23
    personnel changes, 23, 25
    reorganization to Northern Motor, 31–32
    runabout, 24*p*

Index

Northern Motor Car Company
    car reviews, 32–33
    financial problems (1906), 33–34
    models and prices, 33–34, 34*p*
    organization from Northern Manufacturing, 31–32
    purchase by EMF, 68–69, 70

Olds Motor Vehicle Company, 3, 4, 22
Olds, Ransom, 3, 4
Oldsmobile, 17, 37
Owen, Percy, 179

Palms, Charles Louis
    background, 7
    EMF investment, 57
    EMF salary, 58
    officer in EMF, 47
    Wayne investment, 6, 8
Panhard-Levassor, 15
Pelletier, LeRoy
    background, 45–46
    death, 206
    EMF shares, 121
    move to Wayne, 43
    post-Studebaker EMF promotions, 103
    resignation from EMF, 124
    resignation from Flanders Motor, 165
    resignation from Rickenbacker, 202
    Rickenbacker Motor involvement, 198–199
    Studebaker-Flanders 20 promotions, 94, 97
Piquette plant
    assembly consolidation, 41
    purchase by EMF, 134
Pontiac Drop Forge Company, 136
Pontiac Foundry Company, 136
Pontiac Motorcycle Company, 135
Pope, George, 62

Pope Manufacturing Company, 13, 62
Prices of cars
    Columbia Motor, 176
    EMF 30, 52, 65, 72, 123
    Flanders 20, 93, 126
    Northern models, 23, 33
    Rickenbackers, 197, 199, 201
    Studebaker-Flanders 20, 93
    Tiffany electrics, 199
    Wayne models, 8
    Wills Sainte Claire, 180
Production scheduling introduction by Flanders, 40, 41
Puritan Autoparts Company, 175
Puritan Machine Company, 70, 71

Rambler, 38
Red Flag Act (1865), 15–16
Reeves, Alfred, 46, 164
Remington typewriters, 13
Replacement parts business, 70, 71
Rickenbacker, Eddie
    financial success, 196
    four-wheel brakes disappointment, 200
    interest in auto business, 193–194
    resignation from Rickenbacker Motor, 202
Rickenbacker Motor Company
    beginnings, 193–194
    business success (1924), 200–201
    capitalization and officers, 195
    end of business, 202–203
    Everitt's association with, 194, 195
    four-wheel brakes, 199–200
    initial stock sale success, 196–197
    product line and features, 197–198
    recession impact, 201–202

Sachs, Paul, 129
Sanitary Steel Stamping Company, 96

Saxon Motor Car Company, 179
Security Trust Company, 180
Selden court suit, 46
Selling prices of cars. *See* Prices of cars
Severens, Henry, 110, 112
Shanks, Charles, 19
Shaw, John, 137
Shinnick, Donna, 3
Skae, Edward A., 8
Smith, A.O., 135, 137, 140
Smith, Frank, 158
Smith, Paul, 157, 158, 165
Smith, Samuel L., 3
Snell Cycle Fittings Company, 26
Solomon, Sir David, 15
Springfield Metal Body Company, 192
Standard Motor Company, Inc.
   name change to Maxwell Motor, 166
   purchase by United States Motor, 165–166
Stanley steamer, 17
Stanton, Henry, 137
Starley, James, 12
Stearns, Frank, 174
Stevens, Frederick, 122, 129
Stinson Aircraft Company, 184
Stinson, Eddie, 184
Stoddard-Dayton cars, 166
Strong, W.E., 162
Studebaker Brothers Manufacturing Company
   contract with Flanders Manufacturing, 138–139, 140
   EMF and
      approval of EMF production schedule (1910), 123
      cost of loss of EMF, 117
      EMF's new contract proposal, 86–88
      EMF's termination of contract, 102
      reaction to EMF contract concerns, 79–80
      reorganization effects on EMF, 129
      sales contract, 61–62, 64–65, 75–76
      suit against EMF. *See* Studebaker vs. EMF
   end of ties with Flanders, 160
   Everitt and Metzger buyout, 88
   financial problems (1910), 122, 127–128

Studebaker Brothers Manufacturing Company *(continued)*
    Goldman rescue, 128–129
    J.P. Morgan pullout crisis, 128
    J.P. Morgan purchase reports, 118–120
    refusal of Flanders' resignation, 156–157
    relationship with Garford, 62–64, 66
Studebaker, Clement Jr., 90, 122, 137
Studebaker-Flanders 20
    features, 94–95
    pricing plan, 93
    production targets (1910), 97–98
    see also Flanders 20
Studebaker vs. EMF
    circuit court injunction, 110–111
    dismissal of case, 116
    EMF affidavits against Studebaker, 115–116
    EMF injunction rebuttal, 111–112
    injunction overrule, 112
    judge's restrictions on suits, 113
    Studebaker conspiracy revealed, 114
    Studebaker filing of first suit, 108–109
Sullivan, Roger J., 8
Swan, Henry, 109, 111

The Detroit Free Press, 119
The Detroit Journal, 119, 120
The Detroit News, 119, 120
Thomas B. Jeffrey Company, 37
Thornycroft Steamer, 15
Tichenor, Carl, 195
Tiffany Electric Car Company, 199
Toledo Manufacturing Company, 19, 26
Tri-mobile, 136
Trippensee Closed Body Corporation, 199, 201
Trippensee, Frank, 199
Tri-State Sportsman's & Automobile Association, 19
Trout, George, 25
Tucker, Carl, 166
Turnbridge Wells exhibition, 15

United States Motor Company
    component companies, 116, 161
    financial problems (1912), 162
    legal proceedings against, 162–163
    purchase by Standard Motor, 165–166
    reorganization, 163–164

Verville Aircraft Company, 203, 204
Vulcan Gear Works, 136

Wagner Electric Company, 171
Walburn, Thomas S., 39, 42, 44
Walker, Robert, 162
Waverly Electric, 17
Wayne Automobile Company, 60p
    business problems (1907), 10
    capitalization, 8
    Flanders' arrival, 42
    Flanders' production recommendations, 43–44
    management team, 7–8
    product line, 8, 9p
    purchase by EMF, 67–68
    start of, 6
    transformation into EMF, 56–59
Western Malleable Steel & Forge Company, 95, 96
Wheelman's Club, 12
White, Albert, 31
William Gray & Son, 1
Wills, C.H., 180
Wills Sainte Claire Company, 180, 181p
Winternitz & Tauber, 180
Winton, Alexander, 17, 20
Wirth, H.M., 179

Wollering, Max
    EMF exit, 135
    production advances at EMF, 117
    Wayne association, 42, 44
Wright brothers, 16–17

Yale car, 25–26
Younger, Dubois, 184

About the Author

Anthony J. Yanik is editor of *Wheels, Journal of the National Automotive History Collection*, the Detroit Public Library. He has authored numerous articles on automotive history for *Wheels*, as well as for the *Chronicle*, a magazine published by the Historical Society of Michigan, and various other publications. In addition, Mr. Yanik edited the autobiographical notes of Carl Breer that were published by the Society of Automotive Engineers (SAE) under the title *The Birth of Chrysler Corporation and Its Engineering Legacy*. In 1987, he organized and chaired a special historical session of the SAE that featured remarks of five legendary figures in automotive design. He also has presented several papers of his own before the SAE.

Mr. Yanik, now a retiree from General Motors, is a member of The Automotive Press Association, The Society of Automotive Historians, the SAE and its Historical Section, and the Algonquin Club (Michigan/Canadian historians). When not writing, he can be found with watercolor brush in hand and recently had an exhibition of 20 paintings of pre-World War II automobiles.